LAB WORKBOOK

Exploring Agriculture, Food, *and* Natural Resources

Jessica Fife

Hannah Elrick

Publisher
The Goodheart-Willcox Company, Inc.
Tinley Park, IL
www.g-w.com

Copyright © 2024
by
The Goodheart-Willcox Company, Inc.

All rights reserved. No part of this work may be reproduced, stored, or transmitted in any form or by any electronic or mechanical means, including information storage and retrieval systems, without the prior written permission of The Goodheart-Willcox Company, Inc.

ISBN 978-1-68584-883-5

2 3 4 5 6 7 8 9 – 24 – 27 26 25

The Goodheart-Willcox Company, Inc. Brand Disclaimer: Brand names, company names, and illustrations for products and services included in this text are provided for educational purposes only and do not represent or imply endorsement or recommendation by the author or the publisher.

The Goodheart-Willcox Company, Inc. Safety Notice: The reader is expressly advised to carefully read, understand, and apply all safety precautions and warnings described in this book or that might also be indicated in undertaking the activities and exercises described herein to minimize risk of personal injury or injury to others. Common sense and good judgment should also be exercised and applied to help avoid all potential hazards. The reader should always refer to the appropriate manufacturer's technical information, directions, and recommendations; then proceed with care to follow specific equipment operating instructions. The reader should understand these notices and cautions are not exhaustive.

The publisher makes no warranty or representation whatsoever, either expressed or implied, including but not limited to equipment, procedures, and applications described or referred to herein, their quality, performance, merchantability, or fitness for a particular purpose. The publisher assumes no responsibility for any changes, errors, or omissions in this book. The publisher specifically disclaims any liability whatsoever, including any direct, indirect, incidental, consequential, special, or exemplary damages resulting, in whole or in part, from the reader's use or reliance upon the information, instructions, procedures, warnings, cautions, applications, or other matter contained in this book. The publisher assumes no responsibility for the activities of the reader.

The Goodheart-Willcox Company, Inc. Internet Disclaimer: The Internet resources and listings in this Goodheart-Willcox Publisher product are provided solely as a convenience to you. These resources and listings were reviewed at the time of publication to provide you with accurate, safe, and appropriate information. Goodheart-Willcox Publisher has no control over the referenced websites and, due to the dynamic nature of the Internet, is not responsible or liable for the content, products, or performance of links to other websites or resources. Goodheart-Willcox Publisher makes no representation, either expressed or implied, regarding the content of these websites, and such references do not constitute an endorsement or recommendation of the information or content presented. It is your responsibility to take all protective measures to guard against inappropriate content, viruses, or other destructive elements.

Image Credits. Front cover: Darren Baker/Shutterstock.com, luchschenF/Shutterstock.com, Elena Sherengovskaya/Shutterstock.com, BearFotos/Shutterstock.com, Olena Yakobchuk/Shutterstock.com, YuRi Photolife/Shutterstock.com, FFA, Miloje/Shutterstock.com, Volha Hlinskaya/Shutterstock.com

Introduction

This lab workbook is designed for use with the text *Exploring Agriculture, Food, and Natural Resources*. The lessons in the lab workbook correspond to those in the text and should be completed after reading the appropriate text lesson.

Each chapter of the lab workbook contains reviews of the textbook lessons to enhance your understanding of textbook content. The various types of questions include matching, true or false, multiple choice, fill-in-the-blank, and short answer.

Reading *Exploring Agriculture, Food, and Natural Resources* and using this lab workbook will help you acquire a working knowledge of the principles of agriculture, food, and natural resources and their applications. Answering the questions for each chapter will help you master the technical knowledge presented in the text.

The lab workbook lessons also contain activities related to textbook chapter content. The activities range from chapter content reinforcement to real-world application. When you engage in these activities, carefully follow any safety procedures instituted by your teacher.

Contents

UNIT 1
Introducing Agriculture, Food, and Natural Resources

Lesson 1.1 The Wonderful World of Agriculture, Food, and Natural Resources 1
Lesson 1.2 Exploring Opportunities in Agriculture, Food, and Natural Resources 4
Lesson 1.3 Understanding Agricultural Education 7
Lesson 1.4 The History, Mission, and Membership Benefits of the FFA . . . 11
Lesson 1.5 The Supervised Agricultural Experience Program. 14
Lesson 1.6 Proper Procedure for Conducting Meetings 17

UNIT 2
Developing Career Skills in Agriculture

Lesson 2.1 Developing Career Skills in Agriculture 19
Lesson 2.2 Building Leadership Skills through Agriculture 22
Lesson 2.3 Cultivating Attitudes and Habits for Career Success 25
Lesson 2.4 Engaging in Effective Communication 28
Lesson 2.5 Growing through Curiosity and Creativity 30

UNIT 3
The Importance of Agriculture

Lesson 3.1 The Effects of Agriculture on Daily Life. 33
Lesson 3.2 The History of American Agriculture 36
Lesson 3.3 Major US Commercial Crops 39
Lesson 3.4 Agriculture and Climate 42

UNIT 4
Agriculture in Your Community

Lesson 4.1 Agriculture and the United States Economy. 45
Lesson 4.2 The Global Impact of Agriculture 48
Lesson 4.3 Agriculture, Economy, and the Workforce 51

UNIT 5
The Science of Agriculture

Lesson 5.1 Scientists' Work in AFNR 55
Lesson 5.2 Technological Applications in Agriculture, Food, and Natural Resources 59

Lesson 5.3 Emerging Technologies in Agriculture, Food, and Natural Resources 62
Lesson 5.4 Genetic Engineering in Agriculture, Food, and Natural Resources 64

UNIT 6
Skills in Food Science

Lesson 6.1 Introduction to Food Science 67
Lesson 6.2 Food Safety. 70
Lesson 6.3 Principles of Food Preservation. 73
Lesson 6.4 Decoding Food Labels. 76

UNIT 7
Soil Science Exploration

Lesson 7.1 Components of Soil 79
Lesson 7.2 Soil Conservation 82
Lesson 7.3 Soils and Fertilizers 84
Lesson 7.4 Fertilizer Sources. 87

UNIT 8
Plant Science Exploration

Lesson 8.1 What Is Horticulture? 89
Lesson 8.2 Classifying Plants 92
Lesson 8.3 Parts of a Plant 95
Lesson 8.4 Environmental Conditions for Plant Growth. 98
Lesson 8.5 Photosynthesis, Respiration, and Transpiration 101
Lesson 8.6 Reproductive Parts of a Plant 104
Lesson 8.7 Sexual Propagation 107
Lesson 8.8 Asexual Propagation. 110
Lesson 8.9 Agricultural Pests 112

UNIT 9
Animal Science Exploration

Lesson 9.1 Fundamentals of Animal Science. 115
Lesson 9.2 The Importance of the Livestock Industry 118
Lesson 9.3 Major Breeds of Livestock. 121
Lesson 9.4 Introduction to Livestock Selection. 124
Lesson 9.5 Livestock Products and By-Products 127
Lesson 9.6 Livestock Physiology 130
Lesson 9.7 Livestock Feeding and Nutrition 133
Lesson 9.8 Keeping Livestock Healthy 135
Lesson 9.9 Safety around Livestock. 138
Lesson 9.10 Specialty Livestock Production 141
Lesson 9.11 Companion Animals. 143

UNIT 10
Wildlife and Natural Resources Management

Lesson 10.1	Wildlife and Natural Resources Careers	145
Lesson 10.2	Understanding and Researching WFNR Issues	147
Lesson 10.3	Ecosystems	149
Lesson 10.4	Creating Conservation Plans	151
Lesson 10.5	The Wildlife, Forestry, and Natural Resources Industry	153
Lesson 10.6	United States Forest Regions	156
Lesson 10.7	Forests and Forest Products	158
Lesson 10.8	Forest Management Practices	160
Lesson 10.9	Anatomy of a Tree	163
Lesson 10.10	Commercially Important American Trees	165
Lesson 10.11	Natural Resources Management Practices in Agriculture	168
Lesson 10.12	Natural Resources Management Practices in Agriculture	170

UNIT 11
Agricultural Engineering

Lesson 11.1	Safe Agricultural Work Practices	173
Lesson 11.2	Simple Machines and Common Tools	175
Lesson 11.3	Measurement and Layout	181
Lesson 11.4	Common Fasteners and Materials	184
Lesson 11.5	Irrigation and Plumbing	186
Lesson 11.6	Fundamentals of Electricity	188
Lesson 11.7	Small Engine Basics and Maintenance	190
Lesson 11.8	Project Planning, Design, and Construction	194

UNIT 12
Connecting Producers and Consumers through Ag Marketing

Lesson 12.1	What is Marketing?	197
Lesson 12.2	USDA Standards and Grades	200
Lesson 12.3	Determining Price Points	203
Lesson 12.4	Marketing Strategies and Plans	205
Lesson 12.5	Promoting and Advertising Agricultural Products	207
Lesson 12.6	Technology in Marketing and Agribusiness	209

Name _____ Date _____ Class _____

LESSON 1.1 The Wonderful World of Agriculture, Food, and Natural Resources

Lesson Review

Carefully study the lesson and then answer the following questions.

1. Why should everyone care about agriculture? _____
   _____
   _____

2. _____ What does AFNR stand for?
   A. Agriculture, Farming, and Natural Resources
   B. Animals, Flowers, and Nature Resources
   C. Agriculture, Flowers, and Natural Resources
   D. Agriculture, Food, and Natural Resources

3. _____ Hands-on learning is also called _____.
   A. agricultural education
   B. experiential learning
   C. AFNR
   D. career and technical education

4. _____ What is agriculture?
   A. The growing and harvesting of crops and raising domesticated animals for human use
   B. Animals tamed by humans for meat, eggs, dairy, wool, and fibers
   C. Scientists focused on producing efficient and high-quality food and fiber crops while also managing and maintaining the soil
   D. Items originally provided by the natural world, coming from both renewable and nonrenewable resources

5. _____ What does it mean to domesticate?
   A. The growing and harvesting of crops and raising animals for human use
   B. To tame animals for the purpose of producing meat, eggs, dairy, wool, and fiber
   C. To produce efficient and high-quality food and fiber crops while also managing and maintaining the soil
   D. To originate from both renewable and nonrenewable resources

6. Name at least three animals that are considered livestock. _____
   _____
   _____

7. List three products that come from dairy animals such as dairy cows and goats. _____
   _____
   _____

8. _____ What is an agronomist?
   A. The growing and harvesting of crops and raising domesticated animals for human use
   B. An animal tamed by humans for the production of meat, eggs, dairy, wool, and fiber
   C. A scientist focused on producing efficient and high-quality food and fiber crops while also managing and maintaining the soil
   D. An item originally provided by the natural world, coming from renewable or nonrenewable resources

Copyright Goodheart-Willcox Co., Inc.
May not be reproduced or posted to a publicly accessible website.

9. _____ What are natural resources?
   A. The growing and harvesting of crops and raising domesticated animals for human use
   B. Animals tamed by humans for the production of meat, eggs, dairy, wool, and fiber
   C. Scientists focused on producing efficient and high-quality food and fiber crops while also managing and maintaining the soil
   D. Items originally provided by the natural world, coming from both renewable and nonrenewable resources

10. What is an example of a scarce resource on our planet and why is it considered to be scarce? _____

11. How are natural resources and the health of the environment directly tied to the success of agriculture and feeding the world? _____

12. How many pounds of meat are produced from livestock globally each year? _____

13. _____ What are the three components of agricultural education? Select all that apply.
   A. Classroom/laboratory instruction
   B. AFNR
   C. Livestock and natural resources
   D. Leadership (FFA)
   E. Experiential learning (SAE)

14. What national student organization is a part of agricultural education? _____

15. List three AFNR careers that you find interesting and explain why. _____

Name _____ Date _____ Class _____

 **Activity 1.1A**

Recruiting for the Future of AFNR

Congratulations! You've just been hired to recruit fresh talent into the incredible career opportunities within the agriculture, food, and natural resources fields. To encourage your classmates to pursue a job within AFNR, create a sales pitch to present to your class on why they should consider a career in one of these exciting areas! The sales pitch can be for a specific position or for various types of positions within a particular sector (example: poultry industry or water conservation/quality). Don't be afraid to get creative! Use the guide below to get started on your research:

1. What position and/or area are you recruiting for?
   _____

2. What is the job or industry description?
   _____

3. What type of pay range can a potential employee expect?
   _____

4. Does the job or field require workers to have special certification(s), technical, or college degrees?
   _____

5. Are there any big selling points or perks that would be exciting to a potential applicant?
   _____

6. Why did *you* select this position or sector?
   _____

Name _____ Date _____ Class _____

Exploring Opportunities in Agriculture, Food, and Natural Resources

Lesson Review

Carefully study the lesson and then answer the following questions.

1. What are some career options for a student in the Agribusiness Systems career pathway, and what is the typical salary range? _____
   _____
   _____

2. _____ If someone is interested in being a veterinarian, which career pathway should they consider completing?
   A. Power, Structural, and Technical Systems
   B. Natural Resources Systems
   C. Animal Systems
   D. Agribusiness Systems

3. What are some career options for a student in the Environmental Service Systems (or Environmental Science) career pathway, and what is the typical salary range? _____
   _____
   _____

4. What are some career options for a student in the Plant Systems (or Plant Science) career pathway, and what is the typical salary range? _____
   _____
   _____

5. What are some career options for a student in the Natural Resources Systems career pathway, and what is the typical salary range? _____
   _____
   _____

6. _____ If someone is interested in working for the state botanical gardens, which career pathway should they consider completing?
   A. Plant Systems
   B. Natural Resources Systems
   C. Environmental Service Systems
   D. Food Products and Processing Systems

7. What are some career options for a student in the Food Products and Processing Systems (or Food Science) career pathway, and what is the typical salary range? _____
   _____
   _____

8. How would you describe a career pathway? What are three examples of AFNR career pathways? _____
   _____
   _____

Name _____

9. What are some career options for a student in the Biotechnology Systems career pathway, and what is the typical salary range? _____
_____
_____

10. What are some career options for a student in the Animal Systems career pathway, and what is the typical salary range? _____
_____
_____

11. _____ If someone is interested in working for the National Forest Service, which career pathway should they consider completing?
    A. Animal Systems
    B. Food Products and Processing Systems
    C. Agribusiness Systems
    D. Natural Resources Systems

12. What are some career options for a student in the Power, Structural, and Technical Systems (or Agricultural Engineering) career pathway, and what is the typical salary range? _____
_____
_____

Name _____ Date _____ Class _____

Powerful Pathway Presentations

Choose a pathway that interests you and research the many careers that fall within that pathway. Then, research to find an open position within that pathway that you'd like to "apply" for. What are the requirements? What skills will you need to be successful in that position? How can you prepare for that career? Take these questions into consideration as you create a poster presentation that roadmaps the AFNR career pathway you would take, any certifications, technical, or college degrees needed to qualify for the position, and any other skills or experiences you may need to be successful in the job you have selected.

Name _____ Date _____ Class _____

LESSON 1.3 Understanding Agricultural Education

Lesson Review

Carefully study the lesson and then answer the following questions.

1. List three opportunities that agricultural education provides you as a student. _____
   _____
   _____
   _____

2. _____ *True or False?* There are more than 800,000 students participating in agricultural education classes throughout the United States and its territories — just like the one you're in right now!

3. _____ Who is responsible for formalizing agricultural education in 1917?
   A. Dave M. Hughes
   B. Hoke Smith
   C. Dudley M. Hughes
   D. Dave M. Hughes and Hoke Smith
   E. Hoke Smith and Dudley M. Hughes

4. _____ What was the original name of the act that formalized agricultural education?
   A. National Vocational Education Act
   B. Smith-Hughes Act
   C. National FFA Organization
   D. National Association of Agricultural Educators

5. What did the Smith-Hughes Act of 1917 do? _____
   _____
   _____
   _____

6. _____ What did the letters in "FFA" originally stand for?
   A. Fantastic Farmers in Agriculture
   B. Future Farmers of America
   C. Future Farmers and Agriculture
   D. Funny Farmers on Alpacas

7. _____ Which of these activities would be considered "hands-on learning"?
   A. Candling chicken eggs
   B. Planting seeds and watching them grow
   C. Welding
   D. All of these would be considered hands-on learning

8. _____ What are the three components that make up agricultural education?
   A. FFA, SAE, animals
   B. SAE, Smith-Hughes Act, FFA
   C. Classroom/laboratory instruction, FFA, SAE
   D. Leadership, personal growth, career success

Copyright Goodheart-Willcox Co., Inc.
May not be reproduced or posted to a publicly accessible website.

9. Explain the definition and purpose of a Supervised Agricultural Experience (SAE). _____
_____
_____
_____

10. _____ Which of these projects could be considered an SAE?
    A. Managing honeybee hives
    B. Volunteering at a veterinary hospital
    C. Selling vegetables that you grew
    D. All of these could be used as SAE projects

11. _____ Which option best describes the "classroom/laboratory instruction" component of agricultural education?
    A. Personal growth and career success through engagement in FFA programs and activities
    B. Learning by doing
    C. Contextual, inquiry-based instruction and learning through an interactive classroom and laboratory
    D. Experiential service and work-based learning

12. What is another influential piece of legislation that Senator Hoke Smith was involved in before the Smith-Hughes Act? What did it do? _____
_____
_____

13. What is a Career Development Event (CDE)? What is a Leadership Development Event (LDE)? _____
_____
_____

Activity 1.3A

CDE/LDE Exploration

Explore the list of Career Development Events (CDEs) and Leadership Development Events (LDEs) that the National FFA Organization offers for FFA members to compete in. Which one sounds the most exciting to you? Does your chapter compete in this CDE or LDE? What do students learn or do in this contest and what career pathway does it fit into?

Name _____ Date _____ Class _____

Personal Three-Component Model

After reviewing the three-component model of agricultural education, consider what your own personal three-component model would look like. What three components, when combined, would make you your best self? Why? Draw your personal three-component model and share it with your classmates!

Name _____ Date _____ Class _____

LESSON 1.4
The History, Mission, and Membership Benefits of the FFA

Lesson Review

Carefully study the lesson and then answer the following questions.

1. _____ Where and when was the FFA founded?
   A. Kansas City, Missouri in 1928
   B. Kansas City, Kansas in 1928
   C. Indianapolis, Indiana in 1928
   D. Tuskegee, Alabama in 1935

2. _____ Who was elected as the first president of the Future Farmers of America?
   A. George H.W. Bush
   B. Fred McClure
   C. Erwin Milton (E.M.) Tiffany
   D. Leslie Applegate

3. _____ Who was the first black man to serve as a National FFA Officer?
   A. George H.W. Bush
   B. Fred McClure
   C. Erwin Milton (E.M.) Tiffany
   D. Leslie Applegate

4. _____ Who wrote the FFA Creed?
   A. George H.W. Bush
   B. Fred McClure
   C. Erwin Milton (E.M.) Tiffany
   D. Leslie Applegate

5. What was the New Farmers of America organization, and when did the NFA merge with the FFA? _____

6. _____ What are the National FFA colors?
   A. National Blue and Corn Gold
   B. American Blue and Gold
   C. Blue and Yellow
   D. National Blue and Corn Yellow

7. Describe why the National FFA colors were chosen and what they represent. _____

8. What is a creed? _____

Copyright Goodheart-Willcox Co., Inc.
May not be reproduced or posted to a publicly accessible website.

9. _____ How many paragraphs are in the FFA Creed?
   A. 8
   B. 7
   C. 6
   D. 5

10. _____ What phrase starts each paragraph of the FFA Creed?
   A. Agriculture is…
   B. I believe…
   C. We believe…
   D. FFA believes…

11. _____ Who is the "Father of the FFA" and is also credited for designing the FFA emblem?
   A. Leslie Applegate
   B. Erwin Milton (E.M.) Tiffany
   C. Fred McClure
   D. Henry Groseclose

12. Describe the six parts of the FFA Emblem. _____
_____
_____
_____
_____
_____
_____
_____

13. What is a motto? _____
_____

14. What does the FFA motto describe? _____
_____

15. What is the FFA motto? _____
_____

16. What is the FFA mission statement? _____
_____
_____
_____
_____

Activity 1.4A

Building a Personal Emblem, Motto, and Mission Statement

The FFA emblem, motto, and mission statement are all essential components that drive the decisions of the National FFA Organization and its members as a unified group. What if we used the same guidelines for ourselves as individuals and leaders within our school and community? After reviewing the FFA emblem, motto, and mission statement again, create your own emblem, motto, and mission statement that showcase the values that you aspire to live by. Share them with your classmates.

Name _____ Date _____ Class _____

LESSON 1.5: The Supervised Agricultural Experience Program

Lesson Review

Carefully study the lesson and then answer the following questions.

1. What does SAE stand for and how would you describe it? _____
   _____
   _____

2. What are the benefits of having an SAE? List at least three. _____
   _____
   _____

3. _____ What type of SAE is short and introduces students to the skills needed to expand or start a more in-depth SAE that they can carry through their middle and high school careers?
   A. Ownership SAE
   B. School-Based Enterprise SAE
   C. Foundational SAE
   D. Research SAE
   E. Service Learning SAE

4. _____ What type of SAE encourages students to create, own, and operate a business that provides goods and/or services to the marketplace?
   A. Ownership SAE
   B. School-Based Enterprise SAE
   C. Foundational SAE
   D. Research SAE
   E. Service Learning SAE

5. _____ What type of SAE allows students to gain knowledge and understanding in a chosen field as paid employees or unpaid volunteers?
   A. Ownership SAE
   B. School-Based Enterprise SAE
   C. Foundational SAE
   D. Placement SAE
   E. Service Learning SAE

6. _____ What essential part of an SAE involves collecting and entering data?
   A. Data collection
   B. Recordkeeping
   C. Record book
   D. Research

7. _____ In what type of SAE would students choose a research question and work through the scientific method to learn new information or find information that supports existing research?
   A. Ownership SAE
   B. School-Based Enterprise SAE
   C. Foundational SAE
   D. Research SAE
   E. Service Learning SAE

Name _____

8. _____ What type of SAE would be the best fit for a student who enjoys working in partnership with other students, and who would benefit from the school's or FFA chapter's resources?
   A. Ownership SAE
   B. School-Based Enterprise SAE
   C. Foundational SAE
   D. Research SAE
   E. Service Learning SAE

9. _____ What type of SAE would be the best fit for a student interested in volunteering in their community?
   A. Ownership SAE
   B. School-Based Enterprise SAE
   C. Foundational SAE
   D. Research SAE
   E. Service Learning SAE

10. When deciding what type of SAE to start, what four things should you consider?

    A. _____

    B. _____

    C. _____

    D. _____

11. _____ What is another name for an Ownership SAE?
    A. Internship SAE
    B. Entrepreneurship SAE
    C. Research SAE
    D. Agribusiness SAE

12. _____ What is another name for a Placement SAE?
    A. Internship SAE
    B. Research SAE
    C. Recordkeeping
    D. Service Learning SAE

Activity 1.5A

SAE → Dream Job!

Supervised Agricultural Experiences are an excellent way to explore hobbies and interests as career options while gaining experience and employment skills. Let's consider your top three hobbies and how your SAE can connect to those! Once you've listed your favorite hobbies, explore the National FFA Organization's proficiency website (https://www.ffa.org/participate/awards/proficiencies/) to determine if one or multiple hobbies could fall into any of those categories. Your agriculture teacher can assist you with this!

Using the flow chart below (also found on page 53 of the text), figure out what type of SAE would best fit your chosen hobby. Then, make a poster or PowerPoint presentation that proposes your SAE project idea, including how you determined your area, the type of SAE you've chosen, a potential timeline for completion of your SAE, the proficiency area (if applicable), and what you hope to gain from the experience!

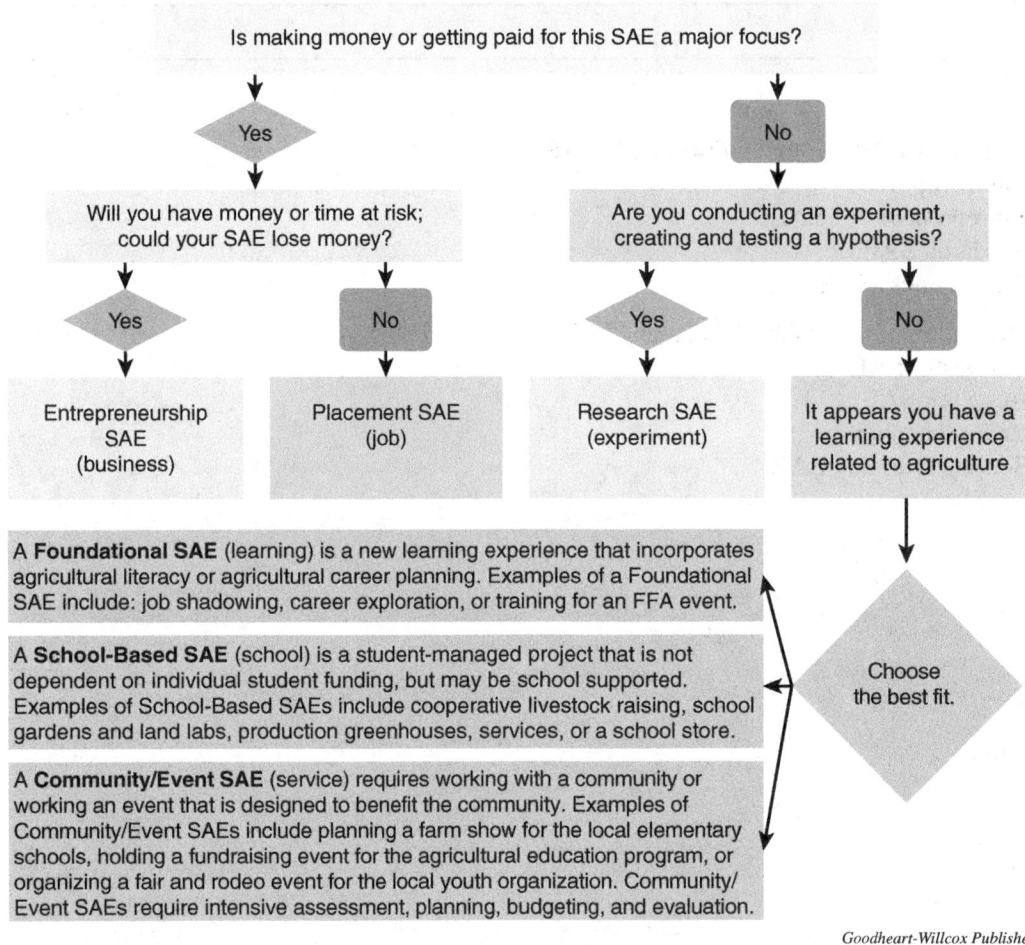

Goodheart-Willcox Publisher

Name _____ Date _____ Class _____

LESSON 1.6
Proper Procedure for Conducting Meetings

Lesson Review

Carefully study the lesson and then answer the following questions.

1. What is parliamentary procedure and why is it used? _____
   _____
   _____

2. _____ The guidelines used in parliamentary procedure come from what source?
   A. *Bob's Rules of Order*
   B. *Applegate's Rules of Order*
   C. *Robert's Rules of Order*
   D. *Richard's Rules of Order*

3. _____ Select the four main rules that ensure meetings run smoothly.
   A. Ensure that decisions are made with the benefit of a large group in mind.
   B. Use of a gavel to keep the room's attention.
   C. Extend courtesy to everyone.
   D. Handle one item at a time.
   E. Be a leader and make decisions for the group.
   F. Ensure that the rights of the smaller population are protected.

Match the terms with the definitions.

4. _____ Subsidiary motion
5. _____ Privileged motion
6. _____ Incidental motion
7. _____ Main motion
8. _____ Order of business
9. _____ Second
10. _____ Quorum
11. _____ Motion
12. _____ Minutes
13. _____ Gavel
14. _____ Chair
15. _____ Amendment

A. The person responsible for conducting meetings
B. The order in which items should be presented in a meeting; also known as the agenda
C. A motion that deals with managing other motions
D. When a second member states that they would also like to discuss a motion with the membership
E. A change to certain aspects of the motion
F. A formal proposal by a member, which, if approved by the membership, will result in action
G. A motion that pertains to the rights and needs of the organization
H. The number of members that must be present in a meeting for formal business to occur
I. A motion that brings a new topic or idea before the membership
J. A motion that relates to parliamentary procedure rules and procedures and not with the business at hand
K. Complete, written record of the events of the meeting
L. A hammer-like tool that the chair uses to direct the membership

Name _____ Date _____ Class _____

 Activity 1.6A

Parliamentary Procedure: Matching for Meetings

Time to get out of your seats and prep for success in parliamentary procedure! By the end of this activity, you'll have an even better understanding of the many details that go into running a successful and smooth meeting using *Robert's Rules of Order*.

1. Split your class into two groups on opposite ends of the room. One group will receive index cards with terminology from this lesson written on them, and the second group will receive index cards with only definitions.
2. At the signal, race to match your card with the corresponding term or definition/phrase.
   A. If you would like to play multiple rounds, shuffle the stacks and redistribute them among your two groups and replay. The repetition will help you retain the information in a fun and interactive way!
3. Once everyone believes they have matched their card correctly, present your cards to the class to review for correctness.

INCREASE THE STAKES: Have your teacher set a stopwatch to keep time. At the end of each round, you'll record the time. There can be a time goal that the class is trying to beat, and if they do so there could be a class reward for their accomplishment.

Terminology and Phrases to Consider Using

Terminology	Phrases
Ballot	Anonymous votes that are written, collected, tallied, and recorded
Roll Call Vote	The chair calls each member by name and asks for his or her vote, one at a time
Rising Vote	The chair asks all those in favor of the motion to stand or raise their hand, then repeat for all those against the motion
Voice Vote	When everyone in favor of the motion responds with an "aye" or "yes," and all those against respond with "nay" or "no"
Striking Out	Allows sections of the motion to be deleted
Inserting	Allows a word or information to be inserted into the motion
Striking Out and Inserting	Allows a portion of the motion to be deleted, and new information to be added to the motion
Amendment	Changing a certain aspect of a motion
Second	When a second member states that they would also like to discuss a motion with the membership
Main Motion	A motion that brings a new topic or idea before the membership
Subsidiary Motion	A motion that deals with managing other motions
Incidental Motion	A motion related to parliamentary procedure rules and procedures and not with the business at hand
Privileged Motion	A motion that pertains to the rights and needs of the organization
Motion	A formal proposal by a member, which, if approved by the membership, will result in a certain action
Minutes	A complete, written record of the events of the meeting
Quorum	The number of members that must be present in a meeting for formal business to occur
Chair	Every meeting must have one of these; the person responsible for conducting the meetings
Order of Business	The order in which items should be presented in a meeting; also called an *agenda*
Gavel	A hammer-like tool that the chair uses to direct the membership
One Gavel Tap	Signifies that the members need to be seated or that the meeting is over
Two Gavel Taps	Signifies the beginning of a meeting
Three Gavel Taps	Signifies that members should stand
Rapid Series of Gavel Taps	Signifies that members should come to order or stop talking
Parliamentary Procedure	Using a set of guidelines called *Robert's Rules of Order* to conduct meetings effectively

Name _____ Date _____ Class _____

LESSON 2.1 Developing Career Skills in Agriculture

Lesson Review

Carefully study the lesson and then answer the following questions.

1. _____ What is character?
   A. A person to look up to or imitate
   B. Having persistence, willingness to fail, and the strength and courage to move forward
   C. The ability to understand and share the feelings or emotions of another person
   D. The attributes and traits that make up and distinguish your individual nature

Match the character trait to a definition or example of the trait.

2. _____ Empathy
3. _____ Gratitude
4. _____ Optimism
5. _____ Grit
6. _____ Role Model
7. _____ Kindness
8. _____ Fairness
9. _____ Honesty
10. _____ Respect

A. Being caring and helpful to others
B. Staying positive in difficult situations
C. Listening attentively when other people speak
D. Determination to try again, even if you have a setback or fail
E. Expressing appreciation to someone
F. Setting an example for your classmates by being friendly to everyone
G. Treating others equally
H. Putting yourself "in someone else's shoes" to understand their feelings
I. Stating the truth when you are asked a question, even if it is difficult

Name _____ Date _____ Class _____

 **Activity 2.1A**

Personal Traits

In your own words, briefly describe what the listed traits mean to you personally. How would you define them? Do you believe these traits are important to display? Why or why not?

Empathy:
_____
_____

Gratitude:
_____
_____

Optimism:
_____
_____

Respect:
_____
_____

Honesty:
_____
_____

Kindness:
_____
_____

Grit:
_____
_____

Fairness:
_____
_____

Is there one trait that you believe is more important than the others listed? Why or why not?
_____
_____

Activity 2.1B

A House of Character

As we learned in Lesson 2.1, *character* refers to the attributes or traits that make up who we are as a person. In Figure 2.1.1 on page 75, an analogy is provided: for people, good character is much like a solid foundation for a house. Character is essential, but developing it requires time and hard work. For this activity, design and draw your "House of Character." What trait(s) make up the foundation of the house that is your personal character? Once you have that, what trait(s) make up the framing or walls of your character? The roof? Does your house have smaller attributes that support your personal character, making up parts of your house like shutters or windows? Present your "House of Character" to a classmate and explain your reasons behind the traits you chose for certain parts of the structure.

Name _____ Date _____ Class _____

LESSON 2.2
Building Leadership Skills through Agriculture

Lesson Review
Carefully study the lesson and then answer the following questions.

1. What is the difference between accountability and responsibility? _____
   _____
   _____

2. Give two examples of positive conflict resolution tactics. _____
   _____
   _____

3. _____ What is *followership*?
   A. Answering for one's actions or decisions
   B. Developing a mission or goals for the group
   C. The ability to follow a leader effectively
   D. Taking responsibility for your own actions

4. _____ What is the exchange of information without words?
   A. Nonverbal communication
   B. Followership
   C. Accountability
   D. Verbal communication

5. _____ What are some examples of personal traits that might make you a better leader? Select all that apply.
   A. Being honest
   B. Working hard
   C. Making sure everyone knows you're the team's leader
   D. Learning from your mistakes

6. Describe a time you were on a team. What went well and what did not? Why do you think that was? _____
   _____

7. _____ What is the use of words to exchange information?
   A. Nonverbal communication
   B. Followership
   C. Accountability
   D. Verbal communication

8. _____ What is the personal quality of being able to guide or direct others?
   A. Team
   B. Leadership
   C. Responsibility
   D. Followership

9. Review the definition of a *short-term goal*. What are two of your short-term goals and how do you plan to accomplish them? _____
   _____
   _____

Name _____

10. Review the definition of a *long-term goal*. What are two of your long-term goals and how do you plan to accomplish them? _____
_____
_____

11. Consider what it means to be a leader. Is there someone in your life that comes to mind when you think of leadership qualities? What made you think of them? Was it a certain event where they showed true characteristics of leadership, or did they demonstrate leadership over time? Describe them and why you see them as a leader.
_____
_____
_____
_____

Name _____ Date _____ Class _____

 Activity 2.2A

Personal Leadership Plan

In this lesson, we learned that a *leader* is someone who guides or directs a group or organization, and a *personal leadership plan* is a strategy for how you will accomplish your goals and grow as a leader. When creating your own personal leadership plan, it's very important to be honest—taking both your strengths and weaknesses into account, including how you can improve upon them to reach your goals. For this activity, we're going to make our own personal leadership plans! To begin, research examples of personal leadership plans online. Once you've decided on your preferred layout, make notes on traits and qualities you possess that fit into your plan. Make a presentation or a poster with your personal leadership plan and present it to your classmates.

Name _____ Date _____ Class _____

LESSON 2.3 / Cultivating Attitudes and Habits for Career Success

Lesson Review

Carefully study the lesson and then answer the following questions.

1. _____ What is the influence that makes you want to do things and finish them, even if they are difficult?
    A. Self-discipline
    B. Motivation
    C. Punctuality
    D. Professionalism

2. _____ What type of motivation speaks to who you are as an individual and what you personally need, want, and value?
    A. Intrinsic
    B. Extrinsic

3. _____ What type of motivation describes a physical reward or validation from another person?
    A. Intrinsic
    B. Extrinsic

4. Write the letter of the term that best describes each example of reward listed.
    A. Intrinsic Reward
    B. Extrinsic Reward

    _____ Clothing

    _____ Shelter

    _____ Job promotion

    _____ Food

    _____ Good grades

    _____ Money

    _____ Praise

    _____ Avoiding penalties

    _____ Improving your skills

5. _____ What is the ability to stay focused on a task or goal without being easily sidetracked?
    A. Self-discipline
    B. Motivation
    C. Punctuality
    D. Time management

6. Why is it important to make a positive first impression? What are some ways you can do this? _____
    _____
    _____
    _____

Copyright Goodheart-Willcox Co., Inc.
May not be reproduced or posted to a publicly accessible website.

7. It takes _____ to make a good first impression, but almost _____ to undo a bad impression!
   _____
   _____

8. What does it mean to be professional? Give three examples. _____
   _____
   _____
   _____

9. _____ What is a process for organizing and planning your day by dividing your time between activities?
   A. Self-discipline
   B. Motivation
   C. Punctuality
   D. Time management

10. What does it mean to have a strong *work ethic*, and what are some qualities of one? _____
    _____
    _____
    _____

11. _____ What is the characteristic of completing a task or meeting an obligation before a previously designated time?
    A. Self-discipline
    B. Motivation
    C. Punctuality
    D. Time Management

Acting Out Professionalism

Now that we know what it takes to be professional and make positive impressions, let's put our knowledge to the test! Your teacher will secretly assign each group a role to act out (i.e. a job candidate and boss in an interview, with one group marked as "professional" and another group as "unprofessional", etc.). Each group will come up with a scripted play or skit to act out in front of the class that demonstrates professional or unprofessional qualities assigned to their role(s). During these short plays, the rest of the class will make notes on the characteristics displayed during each group's play. After the skits, the class should review and discuss their individual notes as a group. Are there any disagreements about what characterizes professional behavior? Have fun with this, but don't forget to be subtle to challenge your classmates!

Name _____ Date _____ Class _____

LESSON 2.4
Engaging in Effective Communication

Lesson Review

Carefully study the lesson and then answer the following by matching the term with the letter of its definition or description.

1. _____ Active Listening
2. _____ Communication
3. _____ Effective Communication
4. _____ Written Communication
5. _____ Feedback
6. _____ Nonverbal Communication
7. _____ Visual Communication
8. _____ Constructive
9. _____ Verbal Communication

A. Helpful to your ability to get better
B. Using body language, facial expressions, and posture to convey information without words
C. Ensures clear, concise, and consistent information is shared, received, and understood by all involved
D. The process of exchanging thoughts, ideas, messages, or information between two or more people
E. A type of verbal communication, as it uses words to convey meaning
F. Information from others about your performance
G. The act of concentrating on the person speaking, understanding their message, and responding appropriately
H. A form of nonverbal communication, like emojis or hieroglyphics
I. Communication using spoken words and sounds

Carefully study the lesson and then answer the following questions.

10. Describe three strategies that can be used to give constructive feedback. _____
_____
_____
_____

11. Describe three effective listening strategies that you would like to better use. Why are these strategies important? How will you work to use them more effectively? _____
_____
_____
_____
_____
_____
_____

# Activity 2.4A

Effective Emoji Expression

Let's test your nonverbal communication skills! Use your phone, laptop, or a sheet of paper to type or draw out a series of emojis that make up a popular phrase, animal name (ex: tiger emoji + shark emoji = tiger shark), or something similar. Once everyone is ready, each person will show their emojis to the class to see how easily they can guess what your message is! Discuss why some emoji phrases were easier to decipher than others. Talk about the details you considered when selecting the emojis you picked to get your message across.

Name _____ Date _____ Class _____

LESSON 2.5 / Growing through Curiosity and Creativity

Lesson Review

Carefully study the lesson and then answer the following questions.

1. _____ What is the capacity to use your imagination to generate or recognize new ideas and possibilities or make something original that is of value?
   A. Process
   B. Creativity
   C. Curiosity
   D. Critical thinking

2. _____ What is information collected through scientific observation?
   A. Scientific method
   B. Critical thinking
   C. Observation
   D. Data

3. _____ What is the process of searching for a solution to a question or problem?
   A. Hypothesis
   B. Critical thinking
   C. Scientific method
   D. Problem-solving

4. _____ What is the ability to think deeply in a clear and informed way?
   A. Process
   B. Critical thinking
   C. Scientific method
   D. Problem-solving

5. _____ What is an informed guess, possible explanation, or answer to your question based on what you already know about a topic?
   A. Hypothesis
   B. Critical thinking
   C. Scientific method
   D. Problem-solving

6. _____ What is a series of actions or steps followed on the way to achieving a goal?
   A. Process
   B. Creativity
   C. Curiosity
   D. Critical thinking

7. _____ What is the strong desire to know or learn something new?
   A. Process
   B. Creativity
   C. Curiosity
   D. Critical thinking

8. _____ What is a process used to gain information through detailed examination and experimentation?
   A. Scientific method
   B. Critical thinking
   C. Observation
   D. Data

Name _____

9. What is the scientific method? _____
_____
_____

10. Number the steps of the scientific method in their correct order.
    A. _____ Test Your Ideas
    B. _____ Ask Questions and Seek Information
    C. _____ Analyze Information and Build Conclusions
    D. _____ Make an Informed Guess

11. List three positive effects creativity can have on your brain. _____
_____
_____

 Activity 2.5A

The Scientific Method

The scientific method is a specific step-by-step process that scientists use to think about and make sense of complex questions. Let's see how well we remember the steps!

Break the class into groups. Each group will receive a set of index cards. On each index card, there will be a step in the scientific method. Racing against other groups in your class, try to put your index cards in the correct order as fast as possible! Then, after shuffling your stack of index cards, swap your stack with another group. The trick? Some stacks are more specific than others and contain more detail! Whichever group has the best average time at the end of the full rotation wins!

Name _____ Date _____ Class _____

LESSON 3.1 The Effects of Agriculture on Daily Life

Lesson Review

Carefully study the lesson and then answer the following questions.

1. What is the difference between a *consumer* and a *producer*? What percentage of the United States population are consumers and what percentage are producers? Why is this important? _____
   _____
   _____
   _____

2. _____ What is a corn-based fuel product that is often added to gasoline?
   A. Cornstarch
   B. Dextrose
   C. Ethanol
   D. Diesel

3. What does USDA stand for? _____
   _____

4. _____ What is the hair-like material that covers sheep?
   A. Wool
   B. Cornstarch
   C. Dextrose
   D. Fiber

5. Which five states produce over one-third of the food Americans eat every day?

   A. _____

   B. _____

   C. _____

   D. _____

   E. _____

6. _____ What is a thread or filament used to create woven or composite material?
   A. Wool
   B. Cornstarch
   C. Dextrose
   D. Fiber

7. What are the top three farm products produced in the United States?

   A. _____

   B. _____

   C. _____

Copyright Goodheart-Willcox Co., Inc.
May not be reproduced or posted to a publicly accessible website.

8. In your own words, why should everyone care about agriculture? Include at least three statistics from the lesson to support your reasoning. _____

_____

_____

9. Using Figure 3.1.4 on page 125, find the consumption per capita for the following commodities:

   A. Fresh and Frozen Fish, 2013: _____ pounds

   B. Ice Cream, 2015: _____ pounds

   C. Fresh Citrus, 2012: _____ pounds

   D. Pork, 2016: _____ pounds

   E. Chicken, 2014: _____ pounds

   F. Corn Products, 2016: _____ pounds

   G. Coffee, 2013: _____ pounds

   H. Peanuts (Shelled), 2015: _____ pounds

Name _____ Date _____ Class _____

 **Activity 3.1A**

Agriculture and Everyday Objects

In this lesson, we've set the foundation for the importance of agriculture and how it affects our daily lives. Let's take it a step further: List ten random objects that you use often or every day. Then, do some research to see how those products are made and where they come from to see if they relate back to agricultural products. Compare your findings with a classmate and pick one product to share with the entire class.

Name _____ Date _____ Class _____

LESSON 3.2
The History of American Agriculture

Lesson Review
Carefully study the lesson and then answer the following questions.

1. What is a boll weevil and how did it wreak havoc on cotton plants? _____
   _____
   _____

2. What caused the Dust Bowl and what detrimental event was taking place at the same time? _____
   _____
   _____

3. What is a plantation? What crops were most common on plantations? In what period were plantations common?
   _____
   _____
   _____

4. _____ Which of the following is *not* a common tool used in precision agriculture?
   A. Global positioning system (GPS)
   B. Drones
   C. Satellite
   D. Steam tractor

5. What did the Pure Food and Drug Act of 1906 do? Why was this law important? _____
   _____
   _____
   _____

6. _____ What invention could remove seeds from cotton fibers?
   A. Cypress McCormick's mechanical reaper
   B. John Deere's steel plow
   C. Eli Whitney's cotton gin
   D. Steam tractor

7. _____ What invention incorporated steel to keep clay in the soil from sticking to the blade, so farmers didn't have to make frequent stops to clean the blades?
   A. Cypress McCormick's mechanical reaper
   B. John Deere's steel plow
   C. Eli Whitney's cotton gin
   D. Steam tractor

8. _____ Who is credited with the invention of barbed wire?
   A. George Washington Carver
   B. John Deere
   C. Joseph Glidden
   D. Eli Whitney

36   Exploring Agriculture Food and Natural Resources Lab Workbook

Name _____

9. _____ What invention resembled a two-wheeled, horse-drawn chariot, with a vibrating cutting blade, a reel to bring the grain within reach, and a platform to receive the falling grain?
    A. Cypress McCormick's mechanical reaper
    B. John Deere's steel plow
    C. Eli Whitney's cotton gin
    D. Steam tractor

10. _____ What is the term for a method of agricultural labor in which the landowner leased farmland to a farmer or tenant in exchange for a portion of the crop produced?
    A. Slavery
    B. Boll weevil
    C. Precision agriculture
    D. Sharecropping

11. What is George Washington Carver known for, and what institution did he direct? _____
_____
_____

12. Consider how American agriculture has changed over the last few centuries. How might life today be different if those advances had not occurred? _____
_____
_____

13. Using Figure 3.2.1 on page 133, name the *continent* on which each food item listed below is believed to have first been domesticated:

    A. Corn: _____

    B. Lettuce: _____

    C. Onion: _____

    D. Orange: _____

    E. Peanut: _____

Name _____ Date _____ Class _____

 Activity 3.2A

The History of Agriculture

In Lesson 3.2, we covered the rich history of the agricultural industry. To give a better idea of some of the biggest stepping stones that helped agriculture progress, break into groups and select a decade, half-century, or century. With your group, research your chosen time frame and select the most important events and individuals (innovations, inventions, societal events, famous scientists) to create a time line. Once every group has completed their section of time line, put them together in chronological order. Each group can then present their findings in turn!

Name _____ Date _____ Class _____

LESSON 3.3 — Major US Commercial Crops

Lesson Review

Carefully study the lesson and then answer the following questions.

1. What are the three types of grain crops?

    A. _____

    B. _____

    C. _____

2. _____ What percentage of the US corn crop is used in animal feed?
    A. 30%
    B. 10%
    C. 48%
    D. 12%

3. What are five different products that can be made from rough rice?

    A. _____

    B. _____

    C. _____

    D. _____

    E. _____

4. _____ What are two examples of a fiber crop?
    A. Cotton, flax
    B. Peanuts, cotton
    C. Cotton, soybeans
    D. Peanuts, flax

Carefully study the lesson and then answer the following by matching the term to its definition or related statistic.

5. _____ Corn
6. _____ Oilseeds
7. _____ Peanuts
8. _____ Grains
9. _____ Cereals
10. _____ Cotton
11. _____ Rice
12. _____ Sorghum
13. _____ Soybeans
14. _____ Flax
15. _____ Wheat

A. The most important fiber crop in the world, accounting for about 35 percent of all fiber crops produced
B. There are four main types of this cereal crop: grain, forage, biomass, and sweet
C. Cereals that are suitable for human consumption
D. This legume is famous in the Southeastern United States, where 99 percent of the entire US crop is grown
E. The fiber of the stem in this crop is used for linen sheets, napkins, tablecloths, clothing, fine paper, and bandages
F. Any grass grown for the edible parts of its grain
G. The most widely produced feed grain in the United States, with more than 80 million acres in production annually
H. This is found in noodles, hard crackers, cereal, and bread
I. The primary food grain for more than half of the world's population
J. Grains that are also valuable for the oil content they produce
K. This legume is one of the most versatile crops grown in the United States and can be found in many things from infant formula to tofu and even candles.

 **Activity 3.3A**

Top Crop Choice

Pick a crop that is grown in the United States. Using this chapter and the internet, research your crop and take detailed notes. Then, make a digital or poster presentation to share with the class. Include statistics, where the crop is grown, how it is harvested, its growing season, its history, and plenty of photos.

Name _____ Date _____ Class _____

LESSON 3.4 Agriculture and Climate

Lesson Review

Carefully study the lesson and then answer the following questions.

1. _____ What absorbs most of the heat produced by the planet?
    A. Greenhouse gases
    B. Ocean
    C. Emissions
    D. EPA

2. _____ What is the measure of current temperature, precipitation, wind, and humidity of an area?
    A. Climate
    B. Weather

3. _____ What does our atmosphere resemble when gases build up in the atmosphere, trapping heat in and warming the planet?
    A. Solar panels
    B. Greenhouse
    C. Water
    D. UV rays

4. _____ What does EPA stand for?
    A. Environmental Progress Agency
    B. Environmental Pattern Association
    C. Environmental Protection Agency
    D. Environmental Pattern Agency

5. What is the EPA's role? _____
   _____

6. _____ When we burn fossil fuels for things like electricity, heat, and transportation, what unintentional by-product do we produce?
    A. Energy
    B. EPA
    C. UV rays
    D. Emissions

7. What is climate change, and what causes it? _____
   _____
   _____

8. Write the percentage of greenhouse gas emissions each economic sector contributes in the blank.

    A. Transportation _____%

    B. Electricity _____%

    C. Industry _____%

    D. Commercial and Residential _____%

    E. Agriculture _____%

42 Exploring Agriculture Food and Natural Resources Lab Workbook

Name _____

9. What are two important ways that climate change affects agriculture? _____
_____
_____

10. Describe ways that agricultural producers and the agriculture industry are trying to address climate change.
_____
_____
_____

11. Describe the difference between *weather* and *climate*. _____
_____
_____

Climate Change Action

Now that we've learned how detrimental climate change can be to our world, let's explore some things the United States and other countries are doing to stop and reverse these negative effects on our environment. Grab a computer and research current laws or legislation that target climate change across the globe. Present your findings to the class. Pay special attention to findings that you and your peers can adopt to make a difference in your community.

Name _____ Date _____ Class _____

LESSON 4.1 Agriculture and the United States Economy

Lesson Review

Carefully study the lesson and then answer the following questions.

1. _____ What is the management and use of money, resources, and labor in a given place?
    A. Economy
    B. Market
    C. Agribusiness
    D. Gross domestic product (GDP)

2. _____ What is the total value of every product created in the United States called?
    A. Economy
    B. Market
    C. Agribusiness
    D. Gross domestic product (GDP)

3. _____ How would you describe items that will begin to spoil and rot if not stored properly and moved to market in a timely manner?
    A. Wholesale
    B. Retail
    C. Perishable
    D. Processing

4. _____ How would you describe foods that are sold to consumers by grocery stores and restaurants?
    A. Wholesale
    B. Retail
    C. Perishable
    D. Processing

5. What is the measurement that equals the movement of 2,000 pounds (one ton) a distance of 5,280 feet (one mile)?

    _____

6. _____ How would you describe foods sold to grocery stores and restaurants for resale?
    A. Wholesale
    B. Retail
    C. Perishable
    D. Processing

7. _____ What process adds value to agricultural products by changing them from raw products to more usable forms?
    A. Wholesale
    B. Retail
    C. Perishable
    D. Processing

8. _____ Which is an example of a processed agricultural product?
    A. Raw milk
    B. Cheese
    C. Fresh-dug peanuts
    D. Fresh-picked watermelons

Carefully study the lesson and then match the questions with the proper amounts.

9. _____ What is the GDP of the United States?
10. _____ How many agricultural jobs are there in the United States?
11. _____ How many food-service related jobs are there in the United States?
12. _____ According to the USDA, what is the value of raw crops, livestock, and livestock products alone in the United States?
13. _____ How many jobs are there in the United States?
14. _____ How much does agriculture contribute to the United States GDP?
15. _____ How many cents of every dollar spent in the United States goes toward purchasing food?

A. 22 million
B. $132 million
C. $21.4 trillion
D. 13
E. 21.6 million
F. 13.5 million
G. $1 trillion

Name _____ Date _____ Class _____

 Activity 4.1A

Commodities and Consumers

Wow, agriculture certainly goes beyond just farming and traditional agricultural practices, doesn't it? Now that we've learned about the economic impact of agriculture in our country, let's zoom in closer to learn more. Pick one commodity, food, or fiber product (raw or processed). Then, using a poster, make a step-by-step flow chart to include the following information:

- Name of agricultural product
- Where is it grown/produced? Why is it grown there (weather, soil, etc. needed)?
- What season is it grown/produced in?
- Does it go through any processing steps? What are the steps and what do they accomplish?
- Is this product sold mostly in a retail or wholesale setting, or both?
- How much money does this product contribute to the US economy?
- Can you tell us how many people work to cultivate and produce the product?

Be sure to use lots of colors and make this eye-catching! Once you've completed your questions and transferred them to your poster, present your commodity to your classmates.

Name _____ Date _____ Class _____

LESSON 4.2 The Global Impact of Agriculture

Lesson Review

Carefully study the lesson and then answer the following questions.

1. How many people in the world go without enough food to eat each day? _____

2. _____ What is the term used for the exchange of desired goods?
   A. Import
   B. Export
   C. Purchasing
   D. Trade

3. _____ To what country does the United States export most of its agricultural products?
   A. Mexico
   B. Kenya
   C. Canada
   D. China

4. What is the condition in which a person does not receive the necessary nutrients in their food to maintain a healthy and well-functioning body? _____

5. _____ What country is the second-largest importer of US agricultural products?
   A. Mexico
   B. Kenya
   C. Canada
   D. China

6. _____ What is the most exported crop of all US agricultural products?
   A. Peanuts
   B. Cotton
   C. Soybeans
   D. Bananas

7. By 2050, how many people will the agriculture industry need to feed, clothe, and shelter? _____

8. _____ What percentage of rice and wheat produced in the United States is shipped abroad?
   A. 50%
   B. 20%
   C. 70%
   D. 10%

9. _____ By how much will global food production need to increase in the next 30 years to sustain the population?
   A. 50%
   B. 20%
   C. 70%
   D. 10%

10. How many cut flowers are imported to the United States each year? _____

11. _____ Which country is *not* one of the largest suppliers of cut flowers that are imported to the US?
    A. Netherlands
    B. Ecuador
    C. Mexico
    D. France
    E. Columbia

Activity 4.2A

Countries and Commodities

In this lesson, we learned that due to limitations in climate, water resources, and suitable farmland, as well as different tastes and cultural desires, no country can produce every agricultural product it needs to sustain itself. For this activity, we're going to explore other countries and what agricultural products they're known for. To begin, select a country you would like to research. Look for reputable sources that provide information on the chosen country's gross domestic product (GDP), their top three agricultural commodities produced, their top trade partners (imports and exports), what agricultural products are being brought in, and what commodities they are exporting. Why is this? Could it be the country's climate? Labor force? What else? Present your findings to your classmates.

 **Activity 4.2B**

Countries and Commodities

Looking to take our first activity to the next level? Each student will receive 10 tokens that represent their top commodities of their country from Activity 4.2A. Based on what was learned from the presentations of your classmates, act as the agricultural trade representative for your country and negotiate trade deals with other students that have products you need and vice versa!

Name _____ Date _____ Class _____

LESSON 4.3
Agriculture, Economy, and the Workforce

Lesson Review

Carefully study the lesson and then answer the following questions.

1. _____ What is the term for a tax on incoming goods from another country?
   A. Supply
   B. Demand
   C. Tariff
   D. Price

2. _____ What refers to the amount of product available?
   A. Supply
   B. Demand
   C. Tariff
   D. Price

3. _____ What refers to the desire that consumers have for a particular product?
   A. Supply
   B. Demand
   C. Tariff
   D. Price

4. _____ What fluctuates based on supply and demand?
   A. Supply
   B. Demand
   C. Tariff
   D. Price

5. _____ According to the lesson, changing demand for which of the following products influenced other segments of the economy?
   A. Corn
   B. Ethanol
   C. Soybeans
   D. Gasoline

6. _____ In 2018, the Pew Research Center reported _____ had the highest average tariff rates, while _____ had the lowest tariff rates in the US.
   A. petroleum/petroleum-based products; clothing and accessories
   B. ethanol; cut flowers
   C. cut flowers; ethanol
   D. clothing and accessories; petroleum/petroleum-based products

7. _____ Why are tariffs used?
   A. To encourage fair trade practices between countries
   B. To allow countries to make more money
   C. To allow farmers to make more money
   D. To make sure prices stay below a certain point

8. In your own words, explain what the law of supply and demand is and how it works. _____
_____
_____
_____

9. What is an example of supply and demand that you have experienced or witnessed in your own life? _____
_____
_____

10. In your own words, explain what an international trade specialist does and what type of skills would be useful or required for someone in this position. _____
_____
_____
_____
_____

Name _____ Date _____ Class _____

 **Activity 4.3A**

The Domino Effect

This lesson showed us how everything is connected to agriculture and how our economy relies on it. The lesson also examined the role that consumers play in agriculture. Draw a web map to demonstrate how a single commodity can have a domino effect on the prices of crops, fiber, livestock, and even by-products such as makeup.

Notes

Name _____ Date _____ Class _____

LESSON 5.1 Scientists' Work in Agriculture, Food, and Natural Resources

Lesson Review

Carefully study the lesson and then answer the following questions.

1. What are the five elements of scientific reasoning?

    A. _____

    B. _____

    C. _____

    D. _____

    E. _____

2. What are four that products George Washington Carver created using peanuts, sweet potatoes, soybeans, and other plants?

    A. _____

    B. _____

    C. _____

    D. _____

3. _____ What is a long period with little or no rainfall called?
    A. Irrigation
    B. Desert
    C. Arid
    D. Drought

4. _____ are scientists who design and build machines and structures.
    _____

5. _____ is the design and construction of machines that do the work of humans.
    _____

6. _____ What is a set of statements that attempts to explain what we see or experience in the natural world?
    A. Hypothesis
    B. Theory
    C. Query
    D. Opinion

7. What is the purpose of precision agriculture?_____
    _____
    _____

8. _____ New ways of farming and ranching to manage agricultural resources responsibly are a part of:
    A. Sustainable agriculture
    B. Precision agriculture
    C. Innovation
    D. Engineering

9. _____ is the study of characteristics that plants and animals inherit from the parent plant or animal.
_____

10. A hypothesis is always written is if it were _____.
_____

11. _____ Who is known as the "Father of the Green Revolution"?
    A. George Washington Carver
    B. Leslie Applegate
    C. Norman Borlaug
    D. Rachel Carson

Name _____ Date _____ Class _____

 **Activity 5.1A**

Chain of Reaction: DDT

Rachel Carson discovered that DDT was harmful to insects and birds. Illustrate a possible food chain showing how the effects of DDT could reach humans. Share your illustration and explain your reasoning to the class.

Two Sides to an Issue: A DDT Discussion

Rachel Carson's book was very controversial. Why do you think that is? Research the benefits of DDT and write an argument in support of it. Then, develop an argument against the use of DDT. Discuss with your classmates. What side of the issue has more support? Where are the arguments stronger?

_____
_____
_____
_____
_____
_____
_____
_____
_____
_____
_____
_____
_____
_____
_____
_____
_____
_____
_____
_____
_____
_____
_____
_____
_____
_____
_____

Name _____ Date _____ Class _____

LESSON 5.2
Technological Applications in Agriculture, Food, and Natural Resources

Lesson Review

Carefully study the lesson and then answer the following questions.

1. _____ What is another term for the application of science, engineering, and mathematics to solve real-world, practical problems?
   A. Regenerative agriculture
   B. Technology
   C. STEM
   D. Urban agriculture

2. _____ _____ is the production of food on vertically aligned surfaces, like stacks of planters along a wall.
   A. Organic farming
   B. Conservation tillage
   C. Urban agriculture
   D. Vertical farming

3. _____ _____ is defined as the production of food that integrates cultural, biological, and mechanical practices that recycle resources, and promote ecological balance while conserving biotechnology.
   A. Organic farming
   B. Conservation tillage
   C. Urban agriculture
   D. Vertical farming

4. What does IPM stand for, what is it, and how is it used? _____
   _____
   _____

5. _____ is a type of light that is invisible to human eyes, but that we feel as heat radiating from objects.
   _____

6. _____ What is the term for an area, typically within a city or town, where the citizens have limited access to affordable and nutritious food?
   A. Urban agriculture
   B. Vertical farming
   C. Food desert
   D. Organic farming

7. _____ _____ is a type of conservation tillage that is combined with other processes to increase biodiversity, reduce soil erosion, and improve the water cycle.
   A. Regenerative agriculture
   B. Pest control
   C. Urban agriculture
   D. Organic farming

8. What did REA stand for, and what was it? _____
   _____
   _____

9. _____ What is another name for the production and distribution of food, fiber, and natural resources in cities and towns?
   A. Organic farming
   B. Urban agriculture
   C. Regenerative agriculture
   D. Vertical farming

10. _____ is the process of preparing soil for planting crops by turning it over and breaking up large chunks to kill weeds and prepare a smooth seedbed for planting.
   _____

11. _____ tillage is a method that involves processes which improve the quality and health of the soil by reducing tillage operations.
   _____

12. _____ What is the term for the amount and variety of life on Earth?
   A. Nature
   B. Regenerative agriculture
   C. Biology
   D. Biodiversity

13. Describe why agricultural literacy is important. _____
   _____
   _____

Activity 5.2A

Agricultural Literacy in Action

There are many labels on food products, some of which use buzzwords or sayings that can be a bit confusing. Find your favorite food product(s) at a local supermarket or online and check out any labels it might have. Do you see a USDA Organic label? Maybe a Non-GMO label, or "No Added Hormones" sticker? Visit the product's website and research any agricultural practices they may utilize. What do those labels mean about the product or ingredients in your product? What regulations, rules, or laws are associated with the labels? Discuss your findings with your classmates.

Name _____ Date _____ Class _____

LESSON 5.3 Emerging Technologies in Agriculture, Food, and Natural Resources

Lesson Review

Carefully study the lesson and then answer the following questions.

1. What does RFID stand for and what does it do? _____
   _____

2. Machines communicating with each other with minimal human interaction are connected through the _____.
   _____

3. The term _____ refers to the ability of computers to "think" and make decisions without human input.
   _____

Circle the correct word in each set of parentheses to complete the following sentence.

4. Agriculture (increased/decreased) the amount of food that humans could grow, but agriculture requires (more/less) effort and skill than hunting and gathering.

5. _____ is the term used to describe the collection and analysis of large amounts of information.
   _____

6. Early agricultural implements were made of _____ and _____.
   _____

7. Archaeologists believe agriculture began in multiple places around the same time. Fill in the blank with the location each crop first appeared using information found in the lesson.

   A. Corn: _____

   B. Potatoes: _____

   C. Pumpkins: _____

   D. Rice: _____

8. What are three examples of data that computers could collect and analyze to improve agriculture?

   A. _____

   B. _____

   C. _____

9. Why would it be beneficial to track the inventory of perishable items, such as fruit and meat, using RFID?
   _____
   _____

10. What are the benefits of auto-steering systems and GPS in modern farm tractors? _____
    _____
    _____

Activity 5.3A

Technology in Agriculture

Just in the last century, the agriculture industry has made technological advances in leaps and bounds. These advances have made food and fiber production exponentially more sustainable and productive. Pick a sector of the agriculture and natural resources industry and research the technologies used in that sector. Select a technology that interests you. Confer as a class to ensure that everyone has selected a unique sector. Continue your research and create a presentation that showcases what your equipment does, and how it impacts agriculture or natural resources. Be sure to incorporate photos or videos of your technology in action!

Name _____ Date _____ Class _____

LESSON 5.4
Genetic Engineering in Agriculture, Food, and Natural Resources

Lesson Review
Carefully study the lesson and then answer the following questions.

1. Famous scientists _____ and _____ are both tied to the development of genetically modified organisms.
   _____

2. _____ GMO zucchini and squash were approved for human consumption in what year?
   A. 1995
   B. 2005
   C. 2015
   D. 1997

3. A(n) _____ is an organism that has had its DNA modified using genetic engineering techniques.
   _____

4. _____ are segments of chromosomes that contain specific information about an organism.
   _____

5. _____ Which GMO crop is both a food and fiber crop?
   A. Beets
   B. Papayas
   C. Alfalfa
   D. Flax

6. _____ What animal was crossed with salmon to produce GMO salmon that grow year-round and reach a marketable weight more quickly?
   A. Sturgeon
   B. Bass
   C. Sea eel
   D. Other salmon species

7. What does DNA stand for and what is it? _____
   _____
   _____

8. _____ What is the process of cutting apart DNA strands and removing or inserting new genes?
   A. Genetic selection
   B. Genetic engineering
   C. Genetic mutation
   D. Genes

9. What are the long threads of DNA called and what do they do? _____
   _____
   _____

Name _____

10. List the GMO crops approved for sale in the US market as of February 2020:

   A. _____
   B. _____
   C. _____
   D. _____
   E. _____
   F. _____
   G. _____
   H. _____
   I. _____
   J. _____
   K. _____
   L. _____
   M. _____
   N. _____
   O. _____

11. Explain how agriculture has benefited from GMOs. _____
_____
_____

Name _____ Date _____ Class _____

 Activity 5.4A

GMOs in the Grocery Store

In this lesson, we learned that there really are not many GMO products available on the shelves of grocery stores. Scan a grocery store website, visit a grocery store in person, or even look through your pantry and refrigerator at home to see how many products have a "non-GMO" label. Make notes of the ingredient list for each product, then compare with the list of GMO products available. How many of those products contain an ingredient that even could be a GMO? Why do you think these labels are used if there isn't a possibility of a GMO product being used?

Name _____ Date _____ Class _____

LESSON 6.1 Introduction to Food Science

Lesson Review

Carefully study the lesson and then answer the following questions.

1. _____ is the study of the physical, biological, and chemical makeup of food, the causes of food deterioration, and the nature of food processing.
   _____

2. _____ _____ occurs when all members of a household always have access to enough food to provide for an active and healthy lifestyle.
   A. Food loss
   B. Food security
   C. Food waste
   D. Food insecurity

3. _____ The Food and Drug Administration is a part of the US Department of _____.
   A. Agriculture
   B. Food and Agriculture
   C. Health and Human Services
   D. Food Science and Development

4. _____ _____ is a term describing the condition experienced by a household that is uncertain of having or unable to acquire enough food to meet the needs of every family member.
   A. Food loss
   B. Food security
   C. Food waste
   D. Food insecurity

5. _____ is a government agency that regulates food quality and nutrition.
   _____

6. _____ _____ includes any edible food that goes uneaten during any stage in the supply chain.
   A. Food loss
   B. Food security
   C. Food waste
   D. Food insecurity

7. The _____ is the combination of people, resources, and processes involved in the production, processing, transportation, and sale of food before it reaches your plate.
   _____

8. _____ refers to high-quality food intended for human consumption that is instead discarded by supermarkets or consumers.
   _____

9. Approximately _____ tons of trash end up in our planet's waterways each year, much of which is plastic, glass, or paper.
   _____

10. How are innovations in food science changing the future? _____
_____
_____

11. What are four tips to reduce food waste?

　　A. _____

　　B. _____

　　C. _____

　　D. _____

 Activity 6.1A

The Plastic Pandemic and the Future of Food

With more than 10 million tons of trash dumped in our planet's waterways each year, it is obvious that single use plastics and waste are detrimental to the health of our planet and ecosystems. The Great Pacific Garbage Patch is a floating island of plastic and trash that has accumulated in the Pacific Ocean due to currents…and is the size of the state of Texas. Microplastics are appearing in human blood, arriving through consumption of affected fish and other seafood. Much of this is attributed to single-use plastics and waste in our lifestyle, but there are strides being made in food science, politics, and new technologies. These can assist in cleaning up the damage that has already been done and in replacing those plastics and garbage that are destroying our environment.

Take some time to explore these issues and research how food science and technology are connected to environmental sustainability and other ways to help reduce food waste. Focus on a topic that you find particularly interesting and present your findings to the class.

Name _____ Date _____ Class _____

LESSON 6.2 Food Safety

Lesson Review

Carefully study the lesson and then answer the following questions.

1. What does CDC stand for? _____
   _____

2. ____ refers to the conditions and practices used to preserve the quality of food to prevent contamination and foodborne illnesses.
   _____

3. A(n) ____ is any illness that is caused by consuming a food that was contaminated with a disease-causing agent such as bacteria, fungi, or another pathogen.
   _____

4. ____ ____ is a harmful bacterium commonly found in the intestines of humans and other animals and can cause severe illness.
   A. Salmonella
   B. E. coli
   C. Bacteria
   D. Fungi

5. What is the best way to prevent the spread of bacteria from one food item to another? _____
   _____

6. Bacterial infections are treated with ____, or drugs that are designed to kill a specific strain of bacteria.
   _____

7. ____ ____ is a harmful foodborne bacterium sometimes found in the intestines of chickens and can be passed on in chicken meat and eggs.
   A. Salmonella
   B. E. coli
   C. Bacteria
   D. Fungi

8. List the Four Cs of Food Safety:

   A. _____

   B. _____

   C. _____

   D. _____

9. Summarize the ideal conditions for harmful bacteria growth. _____
   _____
   _____
   _____

70 Exploring Agriculture Food and Natural Resources Lab Workbook

Name _____

10. _____ What is the safe internal temperature for pork?
   A. 165°F
   B. 145°F
   C. 160°F
   D. Until flesh is pearly or white, and opaque

11. _____ What is the safe internal temperature for shrimp?
   A. 165°F
   B. 145°F
   C. 160°F
   D. Until flesh is pearly or white, and opaque

12. _____ What is the safe internal temperature for chicken?
   A. 165°F
   B. 145°F
   C. 160°F
   D. Until flesh is pearly or white, and opaque

Name _____ Date _____ Class _____

 Activity 6.2A

Partnership for Food Safety Education: Safe Recipe Lesson

Do you think you can use your knowledge of food safety to help you prepare a recipe safely? Let's find out!

Using a device with internet access, navigate to a search engine and search for "Partnership for Food Safety Education safe recipe lesson." Scroll to make sure you find the right link—it should be from fightbac.org. You will find a fun activity where you can help fill in basic instructions for safe food handling, preparation, and storage.

When you've finished the activity, come back for some reflection. Which elements of the activity were easy? Which elements were hard? What lessons can you take away to help you keep your food safe at home?

_____
_____
_____
_____
_____
_____
_____
_____
_____
_____
_____
_____
_____
_____
_____
_____
_____
_____
_____
_____
_____
_____
_____
_____
_____
_____

Name _____ Date _____ Class _____

LESSON 6.3 Principles of Food Preservation

Lesson Review

Carefully study the lesson and then answer the following questions.

1. The process of _____ involves heating food to a temperature of approximately 240 degrees for at least 15 minutes.
   _____

2. _____ is the amount of time food stays good.
   _____

3. _____ What is the science of maintaining the form, health, and nutrition of our edible food?
   A. Food science
   B. Food preservation
   C. Preservatives
   D. Additives

4. _____ uses hot air to preserve food.
   _____

5. _____ _____ are chemicals added to foods to inhibit the growth of microbes.
   A. Preservatives
   B. Fermentation
   C. Picklings
   D. Additives

6. _____ uses heat and pressure to raise the temperature of foods above 212 degrees and provide an airtight seal around the lids of cans or glass jars.
   _____

7. _____ involves first freezing food and then removing the moisture under pressure.
   _____

8. _____ is the process of food breaking down to the point where it is no longer healthy or safe to eat.
   _____

9. _____ What process is also referred to as *pickling* and uses helpful microorganisms or bacteria to slow the growth of harmful bacteria?
   A. Canning
   B. Food preservation
   C. Fermentation
   D. Additives

10. By reducing the temperature of foods to between _____ and _____ degrees Fahrenheit, we can significantly extend their shelf life.
    _____
    _____

Copyright Goodheart-Willcox Co., Inc.
May not be reproduced or posted to a publicly accessible website.

11. Describe how preserving food products protects our food supply. _____
_____
_____
_____

12. Identify factors that contribute to the deterioration of food. _____
_____
_____
_____

13. List basic food preservation methods and techniques.

   A. _____
   B. _____
   C. _____
   D. _____
   E. _____
   F. _____

Name _____ Date _____ Class _____

 **Activity 6.3A**

Food Preservation Technique Identification

Can you identify preserved foods? Collect at least two images of products that fall into each category of food preservation that you learned about in Lesson 6.3. This activity can be done in different ways, depending on available time and resources. You may use a grocery store website where products can be viewed online, or you may take pictures of food in your own pantry and refrigerator at home, or you can take pictures at a local store.

How did you determine how those foods were preserved? Could any of your products be categorized in multiple ways? How? Discuss your findings with your classmates.

Name _____ Date _____ Class _____

LESSON 6.4
Decoding Food Labels

Lesson Review

Carefully study the lesson and then answer the following questions.

1. Why are nutrition labels important? _____
   _____
   _____

2. _____ According to the Dietary Guidelines for Americans, added sugars should be limited to _____ of total daily calories.
   A. 10 percent
   B. 5 percent
   C. 15 percent
   D. 1 percent

3. A(n) _____ is a standard amount a person should eat at one time.
   _____

4. _____ is a reference that tells us how much of each nutrient can be found in one serving of the labeled food.
   _____

5. _____ is the energy produced by the food when burned or metabolized by the body.
   _____

6. _____ provide basic information about the contents and healthfulness of the food contained in the attached package.
   _____

7. _____ are individual foods combined to make a food product or recipe.
   _____

8. _____ What is the recommended total daily caloric intake for most adults?
   A. 1,000 calories
   B. 1,500 calories
   C. 2,000 calories
   D. 2,500 calories

9. _____ When were food nutrition labels introduced in the United States?
   A. 1980s
   B. 1960s
   C. 1990s
   D. 1970s

10. The _____ was passed in 1990 and required food manufacturers to provide a detailed Nutrition Facts label with standard information on every food item produced.
    _____

Name _____

11. Name three trusted sources for nutrition information.

   A. _____

   B. _____

   C. _____

12. _____ Who regulates all labels and language that may be used on food?
    A. United States Department of Agriculture
    B. United States Food and Drug Administration
    C. United States Department of Food and Agriculture
    D. The companies that produce the food products

Name _____ Date _____ Class _____

 Activity 6.4A

Fast Food Nutrition Facts

Search online for Nutrition Facts labels for your favorite fast food restaurant meal and answer the following questions:

Restaurant name: _____

Meal name and/or list of items typically ordered: _____
_____
_____

What is the portion size of the meal? _____

How much fat is in the entire meal? Include trans and saturated fats. _____

How much sugar is in the entire meal? _____

How many calories are in the entire meal? _____

Do your findings surprise you? Why? _____
_____
_____
_____
_____

How might you adjust your order to make it a more balanced meal? _____
_____
_____
_____

Name _____ Date _____ Class _____

Components of Soil

Lesson Review

Carefully study the lesson and then answer the following questions.

1. _____ What layer of the soil horizon is also known as the subsoil?
   A. B
   B. A
   C. O
   D. R
   E. C

2. _____ is a combination of minerals, gases, liquids, living organisms, and organic matter that supports the growth and development of plants and animals.
   _____

3. _____ What tiny organisms are found in great abundance in the soil?
   A. Nematodes
   B. Bacteria
   C. Protozoa
   D. Algae

4. There are (more/the same number of/fewer) living organisms in a handful of soil than there are people on Earth.
   _____

5. _____ Which of the following is in the correct order of soil particles from largest to smallest?
   A. Clay, silt, sand
   B. Sand, silt, clay
   C. Silt, sand, clay

6. _____ is another word for organic matter.
   _____

7. _____ _____ are multicellular organisms that resemble worms and feed on bacteria.
   A. Algae
   B. Protozoa
   C. Arthropods
   D. Nematodes

8. _____ What soil organism plays an important role in producing organic matter?
   A. Protozoa
   B. Nematodes
   C. Algae
   D. Bacteria

9. The arrangement of horizons in the soil is called the _____.
   _____

10. _____ are single-celled organisms that feed on bacteria.
    _____

11. What three types of characteristics distinguish one soil horizon from another?

    A. _____

    B. _____

    C. _____

12. _____ is soil made with equal parts sand, silt, and clay.

_____

13. Draw and label a pie graph that represents the soil components of an ideal soil.

Name _____ Date _____ Class _____

 **Activity 7.1A**

SoilWeb: An Online Soil Survey Browser

The SoilWeb map allows you to explore the US Department of Agriculture National Cooperative Soil Survey data for locations throughout most of the United States. You can use your phone, laptop, desktop computer, or a tablet for this activity.

To begin, log on to your internet-connected device and go to a search engine. Search for "SoilWeb" and choose the University of California-Davis result. Select "SoilWeb" as the site you'd like to visit, then choose a location you'd like to research — perhaps your house, favorite park, or somewhere you went on vacation. Once you've zoomed in on your chosen location, click on the specific area you'd like to look at (like a backyard), and check out the soil information provided in the top left corner of your screen. What do these terms mean about the soil in your location? Share your findings with a classmate.

Name _____ Date _____ Class _____

LESSON 7.2
Soil Conservation

Lesson Review

Carefully study the lesson and then answer the following questions.

1. _____ is another term for the healthy variety of organisms that exist within the soil, including bacteria, fungi, and earthworms.
   _____

2. _____ What major river contributes approximately 20 percent of the soil deposition that enters oceans around the world?
   A. Nile
   B. Amazon
   C. Mississippi
   D. Maumee

3. _____ Landslides are examples of _____.
   A. Soil deposition
   B. Soil detachment
   C. Soil transport

4. List the three steps of soil erosion.

   A. _____

   B. _____

   C. _____

5. _____ Erosion can cause problems for which of the following? Circle all that apply.
   A. Wildlife
   B. Water Quality
   C. Roadways
   D. Buildings and Structures

6. List five methods of preventing soil erosion.

   A. _____   D. _____

   B. _____   E. _____

   C. _____

7. NRCS stands for _____.
   _____

8. Erosion (increases/decreases/does not affect) the amounts of air, water, and nutrients available in the soil.
   _____

9. _____ is a method of farming that prevents the erosion of soil.
   _____

10. What are the two most common causes of erosion?

    A. _____   B. _____

Exploring Agriculture Food and Natural Resources Lab Workbook

Copyright Goodheart-Willcox Co., Inc.
May not be reproduced or posted to a publicly accessible website.

Name _____ Date _____ Class _____

 **Activity 7.2A**

I Spy: Erosion Edition

Review Lesson 7.2, then take a walk around your school grounds or home. Look for places experiencing erosion or erosion prevention methods that are being implemented and take a photo. Answer the following questions for each spot that you find:

Spot 1

Location: _____

What do you think could be causing the erosion issues? _____
_____
_____
_____

Are there methods being used to prevent erosion? If not, what could be a possible solution to help? _____
_____
_____
_____

Spot 2

Location: _____

What do you think could be causing the erosion issues? _____
_____
_____
_____

Are there methods being used to prevent erosion? If not, what could be a possible solution to help? _____
_____
_____
_____

Once everyone is back together, share your results with your classmates and discuss.

Name _____ Date _____ Class _____

LESSON 7.3
Soils and Fertilizers

Lesson Review

Carefully study the lesson and then answer the following questions.

1. List the eight micronutrients:

    A. _____   E. _____

    B. _____   F. _____

    C. _____   G. _____

    D. _____   H. _____

2. _____ calculate the nutrient levels available in the soil.
    _____

3. Micronutrients are needed in _____ amounts that are essential for plant growth and development.
    _____

4. _____ The three elements labeled on a bag of fertilizer are:
    A. Ca-N-K
    B. N-P-K
    C. M-P-N
    D. M-P-K

5. _____ Which option is *not* a fertilizer application method?
    A. Broadcast
    B. Starter solutions
    C. Foliar feed
    D. pH test

6. A(n) _____ is missing one or more of the primary macronutrients and a packaging label would indicate _____ for that macronutrient.
    _____

7. _____ is a measurement of the hydrogen activity in soil.
    _____

8. _____ are nutrients that plants need in large quantities.
    _____

9. A(n) _____ is a substance that is spread on the ground or mixed in the soil to provide nutrients for plant growth and health.
    _____

10. Discuss factors that affect soil pH. _____
    _____
    _____

Name _____

11. Why does pH matter? _____
_____
_____

12. What is the difference between primary macronutrients and secondary macronutrients? _____
_____
_____

13. List three secondary macronutrients.

    A. _____

    B. _____

    C. _____

14. What are the jobs of macronutrients and micronutrients? _____
_____
_____

Name _____ Date _____ Class _____

 Activity 7.3A

Homemade pH Testing Activity

After reviewing Lesson 7.3, gather materials needed to collect small soil samples across the school grounds or at home, such as small cups, a small shovel or spade, marker, etc.

You'll collect two samples from each location that you choose.

Write your soil collection locations on your cups and collect some small samples. Then, complete the following steps:

1. Place 2 tablespoons of soil from one location in a bowl and add ½ cup vinegar. If the mixture fizzes, you have alkaline soil.

2. Place 2 tablespoons of soil from the same location in a bowl and moisten it with distilled water. Add ½ cup baking soda. If the mixture fizzes, you have acidic soil.

Take notes on how your soil samples from each location react. Compare your findings with classmates if you completed this activity at home.

Name _____ Date _____ Class _____

Fertilizer Sources

LESSON 7.4

Lesson Review

Carefully study the lesson and then answer the following questions.

1. A(n) _____ is a naturally occurring nutrient material that originates from plants or animals.
   _____

2. _____ (also known as *synthetic fertilizers*) are nutrient compounds that are derived from materials other than plants or animals.
   _____

3. _____ _____ is decomposed organic matter that can be added to the soil to help plants grow.
   A. Manure
   B. Organic fertilizer
   C. Inorganic fertilizer
   D. Compost

4. _____ are a combination of animal feces and bedding material, such as straw or wood shavings.
   _____

5. _____ A(n) _____ is a secondary product made in the manufacturing of something else.
   A. manure
   B. organic fertilizer
   C. by-product
   D. inorganic fertilizer

6. Ammonia was discovered in 1908 by German chemists _____ and _____.
   _____

7. _____ Native Americans used _____ as a fertilizer.
   A. manure
   B. compost
   C. fish
   D. bison

8. Successful composting requires: _____
   _____
   _____

9. _____ Which of the following are solid nitrogen fertilizers? Circle all that apply.
   A. Ammonium sulfate
   B. Ammonium nitrate
   C. Ammonium phosphate
   D. Urea

10. _____ What does *organic* mean, as it relates to fertilizer?
    A. Healthier for you
    B. No fertilizers or chemicals used
    C. Materials derived from living matter
    D. Some fertilizers or chemicals used

Copyright Goodheart-Willcox Co., Inc.
May not be reproduced or posted to a publicly accessible website.

Name _____ Date _____ Class _____

Activity 7.4A

Environmental Lunch Log Worksheet

For three days, fill out the log below at lunchtime, noting what you eat and what you do with your trash. Review your results with the class. Are you surprised at the results in any of the columns? How can we improve our habits to create healthier soils and compost materials at our schools and homes?

Item Description	Reuse	Recycle	Compost	Landfill	Could Replace With
Banana peel			X		
Sandwich bag	X				
Juice box				X	Reusable thermos or reusable plastic bottle of juice

Name _____ Date _____ Class _____

LESSON 8.1 What Is Horticulture?

Lesson Review

Carefully study the lesson and then answer the following questions.

1. What are the two main types of horticulture? _____
   _____

2. _____ Which of the following is *not* a form of ornamental horticulture?
   A. Pomology
   B. Floriculture
   C. Landscape and nursery production
   D. Turf grass

3. What are the benefits of interiorscaping? _____
   _____
   _____
   _____

4. What is a nursery? _____
   _____
   _____
   _____

5. _____ Which of these countries contains some of the largest greenhouses in the world?
   A. Eritrea
   B. The United States
   C. Spain
   D. The Netherlands

6. _____ Turf grasses are grown for which of the following applications? Circle all that apply.
   A. Golf courses
   B. Lawns
   C. Gardens
   D. Athletic fields

7. _____ Which of these is *not* a main component of food production horticulture?
   A. Floriculture
   B. Olericulture
   C. Viticulture
   D. Pomology

8. What parts of a plant can produce a vegetable? _____
   _____
   _____

9. What are some of the most popular pomology crops grown in the United States? List as many as you can.

_____

_____

_____

10. Why are grapes grown on trellises? _____

_____

_____

_____

11. Horticulture is defined in this lesson as the production, processing, and sale of plants for food, comfort, and beauty. Give an example of a crop grown for each of these purposes. _____

_____

_____

_____

Name _____ Date _____ Class _____

 **Activity 8.1A**

Growing a Horticulture Career

Horticulture is a broad and diverse field, covering numerous branches. Make a list of as many horticulture-related careers as you can think of. Split your list into two categories: Ornamental Horticulture (Floriculture, Landscaping and Nursery Production, Interiorscaping, Turf Grass) and Food Production Horticulture (Olericulture, Pomology, Viticulture). Do not be afraid to use the internet to do some more digging on the topic! Sites like FFA and Seed Your Future may help.

Once you have a full list, choose one career to focus on. What about this career is interesting to you? What training do you need to get this job? How many people have this career? What skills would you need? What would be difficult about it? What would be fun about it? Share your thoughts with the class.

Ornamental Horticulture (Floriculture, Landscaping and Nursery Production, Interiorscaping, Turf Grass)

_____
_____
_____
_____
_____
_____
_____
_____
_____
_____
_____
_____
_____

Food Production Horticulture (Olericulture, Pomology, Viticulture)

_____
_____
_____
_____
_____
_____
_____
_____
_____
_____
_____
_____
_____

Name _____ Date _____ Class _____

LESSON 8.2 Classifying Plants

Lesson Review

Carefully study the lesson and then answer the following questions.

1. In your own words, how would you explain the importance of scientific names? _____
   _____
   _____
   _____

2. A(n) _____ is a group of plants from the same species grown for their desirable, reproduceable traits.
   _____

3. Who created the standard for scientific naming that is followed today? _____
   _____

4. _____ While there are four classifications of plants, the two most commonly found in the horticulture industry are:
   A. Gymnosperms and monocots
   B. Monocots and dicots
   C. Angiosperms and gymnosperms
   D. Angiosperms and shrubs

5. Are there more species of monocots or dicots? _____
   _____

6. What are the different characteristics that are used to classify plants? Name at least three. _____
   _____
   _____
   _____
   _____
   _____

7. What does the USDA Plant Hardiness Zone Map tell gardeners, and how would you use it? _____
   _____
   _____
   _____
   _____

8. _____ refers to whether plants lose their leaves in the fall or keep them all year.
   _____

Name _____

9. _____ What type of plant cycles through its life cycle in a single season or year?
   A. Perennial
   B. Annual
   C. Deciduous
   D. Biennial

10. _____ Which of these plants are biennials? Circle all that apply.
    A. Celery
    B. Asparagus
    C. Hollyhocks
    D. Garlic

11. Most _____ perennials die back in the winter months before returning in the spring.

_____

Name _____ Date _____ Class _____

Activity 8.2A

Fun with the USDA Zone Hardiness Map

If you go online and find the Zone Hardiness Map on the US Department of Agriculture's website, you will see that the map covers a large number of zones, from 1a to 13b. Looking at the map, answer these questions:

1. What zone do you live in? What are three other cities or towns outside of your state that share that same zone classification?

   _____

   _____

   _____

2. Where are the coldest zones located?

   _____

   _____

3. Where are the warmest zones located?

   _____

   _____

Now, choose two zones: one cold (numbered 1–4) and one warm (numbered 9–13). Using online resources, make a list of 10–12 plants that grow in each of these zones. Then, answer these questions:

1. What do the plants in each of these zones have in common?

   _____

   _____

   _____

2. What differences do you see between the plants across zones?

   _____

   _____

   _____

Name _____ Date _____ Class _____

LESSON 8.3
Parts of a Plant

Lesson Review

Carefully study the lesson and then answer the following questions.

1. Name the six main parts of a plant.

   A. _____   D. _____

   B. _____   E. _____

   C. _____   F. _____

2. ____, ____, and ____ are the vegetative parts of a plant.

   _____

   _____

   _____

3. _____ Most nutrient absorption occurs through the:
   A. Seeds
   B. Roots
   C. Root hairs
   D. Xylem

4. Monocots mostly have _____ roots.

   _____

5. What are the three main external parts of the stem, and what do they do? _____

   _____

   _____

   _____

6. What is the difference between xylem and phloem? _____

   _____

   _____

   _____

7. Name five different plants with modified stems that serve as underground food and water storage organs.

   _____

   _____

   _____

   _____

   _____

8. Why are leaves important to a plant? _____
   _____
   _____
   _____

9. What is the only fruit that has seeds on the outside? _____
   _____

10. _____ _____ act as protective tissue around seeds.
    A. Corms
    B. Fruits
    C. Monocots
    D. Embryos

Name _____ Date _____ Class _____

 **Activity 8.3A**

Help Wanted: Plant Parts

These plant parts are looking for a job. Can you help place them in the job opening for which they are best suited?

A. petals
B. stamen
C. phloem
D. pistil
E. root cap
F. chloroplasts
G. sepals
H. taproot
I. xylem

Job openings:

1. _____ Advertising Executive: Colorful personality needed to advertise availability of pollen and nectar.

2. _____ Waitstaff: Deliver food to hungry plant cells. Work in busy roots, stems, and leaves.

3. _____ Drilling Specialist: If you have experience digging deep in search of water, we're looking for you. No branching necessary!

4. _____ Personal Security: Help needed to protect buds. Apply before spring!

5. _____ Hydraulic Engineer: Can you carry some water? Start at the roots and work your way up to the top!

6. _____ Construction Worker: Drill for water, and protect other members of water search team. Apply at root tip!

7. _____ Farmer: Help needed to manage egg production and receive pollen.

8. _____ Chemist: Can you convert carbon dioxide and water into glucose? You must wear green and enjoy working in the sun.

9. _____ Pollen Production Assistant: Help needed to produce pollen! Seasonal work only.

Name _____ Date _____ Class _____

LESSON 8.4 Environmental Conditions for Plant Growth

Lesson Review

Carefully study the lesson and then answer the following questions.

1. List three reasons that light is critical for plants.

    A. _____

    B. _____

    C. _____

2. Every plant species has a preferred _____ that will lead it to its best growth.
   _____

3. When are plants that require a lot of sunlight to flower most likely to bloom? Why? _____
   _____
   _____

4. What is a sign that a plant is not getting enough light? _____
   _____

5. _____ *True or False?* Plants that require low amounts of light will bend away from a light source.

6. _____ Most plants grow well in the temperature range between:
    A. 20° and 40° F
    B. 20° and 40° C
    C. 50° and 85° C
    D. 50° and 85° F

7. How can growers manage greenhouse temperatures and improve plant growth? List at least three methods.
   _____
   _____
   _____

8. Why do plants grow well in areas with reasonably high humidity? _____
   _____
   _____

9. Why is it important to monitor the air quality in a greenhouse? _____
   _____
   _____
   _____

Name _____

10. Water makes up between ____ and ____ percent of horticultural plants.
_____

11. How can you tell that plants are getting inappropriate amounts of water? _____
_____
_____
_____

12. Why do growers use different types of irrigation? What considerations might influence a grower to choose a particular irrigation type? _____
_____
_____
_____
_____

Name _____ Date _____ Class _____

 Activity 8.4A

Greenhouse Watering Lab

Do you often find your plants are wilting from lack of water? Or maybe you have the opposite problem: Your plants are drowning from too much water and turning yellow or having algae problems. It can prove quite difficult to water your plants properly! Knowing the water needs of each plant that you are trying to grow is crucial for the health of your plants. It is also important to water evenly and correctly so that some plants do not dry out faster than others.

Supplies

- 606 Cell pack tray inside a webbed tray (or similar)
- Premoistened soil
- Hard plastic clear cups (must be durable to support weight of the tray)
- Watering wand or watering can

Instructions

1. Fill the entire cell pack tray with premoistened soil.
2. Arrange 8–10 hard plastic cups on a flat surface and place the tray on top of them.
   - Spread the cups out evenly, making sure that the corners and center of the tray are supported.
3. Use a watering wand to evenly water the entire tray. Stop once you think you have the entire tray watered evenly.
4. Remove the tray from the cups and place it to the side.
5. Record your observations!

Questions

1. What did you observe when you looked at the different cups? How many of the cups had the same amount of water?

   _____
   _____
   _____

2. Did the cups along the edge of the tray contain more or less water than the ones in the center? Why do you think this is?

   _____
   _____
   _____

3. How much water was in each cup? Why do you think the soil didn't absorb all the water?

   _____
   _____
   _____

4. What steps should you take to improve your ability to water evenly?

   _____
   _____
   _____

Name _____ Date _____ Class _____

LESSON 8.5
Photosynthesis, Respiration, and Transpiration

Lesson Review

Carefully study the lesson and then answer the following questions.

1. Plants manufacture food through a process called _____.
   _____

2. What does the process of photosynthesis look like, when written as an equation?
   _____

3. _____ Which organisms produce the most oxygen?
   A. Phytoplankton
   B. Trees
   C. Garden shrubs
   D. All land plants

4. Why is sunlight necessary for photosynthesis to take place? _____
   _____
   _____

5. What part of the plant is comparable to humans' lungs? _____
   _____

6. Respiration produces energy for what purpose? _____
   _____
   _____

7. _____ *True or False?* Temperature is related to the rate of respiration, which dictates how quickly plants grow.

8. _____ When does respiration primarily occur?
   A. In the morning
   B. At night
   C. During daylight hours
   D. Throughout the day

9. Why is transpiration important to plant growth and health? _____
   _____
   _____
   _____

10. How do you think that cacti benefit from not transpiring? _____
    _____
    _____

Copyright Goodheart-Willcox Co., Inc.
May not be reproduced or posted to a publicly accessible website.

Name _____ Date _____ Class _____

 Activity 8.5A

Do Plants Consume or Release Carbon Dioxide?

You have probably heard that plants do a lot of work to absorb carbon dioxide (CO_2) from our atmosphere. However, when you look at the processes of photosynthesis and respiration in this chapter, you may have noticed that one absorbs CO_2 (photosynthesis) and one gives off CO_2 (respiration). Do plants absorb or give off more CO_2? Let's find out!

This activity will determine whether CO_2 is consumed or produced as elodea plants are placed in dim light, bright light, or a dark environment. The change in CO_2 will be detected by the pH indicator phenol red. Phenol red is yellow under acidic conditions (high hydrogen ion concentrations), pink under basic or alkaline conditions (low hydrogen ion concentrations), and orange under neutral conditions. If the elodea plants are releasing oxygen into the water or absorbing oxygen from the water, this will change the hydrogen ion levels in the water, resulting in a visible color change.

- If the CO_2 concentration decreases, the H ion concentration will also decrease and the solution will change to pink, becoming basic.
- If the CO_2 concentration increases, the H ion concentration will also increase and the solution will change to yellow, becoming acidic.
- Neutral solutions of phenol red will be orange.

Supplies
- Aluminum foil
- Elodea (Anacharis) plants
- 100ml Graduated cylinder
- Parafilm
- Phenol red
- Straw
- 6 Test tubes
- Water

Instructions

1. Create a solution of phenol red by adding concentrated phenol red to 100 ml of water. The phenol red may change color as a result of adding water (depending on the acidity of the water). Your goal is to make your solution a neutral orange color. You can do this by gently blowing into the solution with a straw in small amounts at a time.

2. Once you have an orange solution, transfer it into the six test tubes until each test tube is two-thirds full.

3. Cut 3 pieces of the elodea plant about 2 inches in length. Place a cut piece of elodea (cut end facing up) into 3 of the 6 tubes. The other 3 test tubes will not have elodea and will serve as controls.

4. Cut off 6 pieces of parafilm. Each piece should be about 1 ½ inches square. Cover each test tube with a piece of parafilm. Gently pull the parafilm over the opening of the test tube until there is a tight seal. Rub your finger along the edge of the rim to ensure there are no air bubbles under the seal.

5. Tear off 2 pieces of aluminum foil that are about 6 inches long. Select one test tube containing the elodea plant and select a test tube with no plant. Wrap each of these test tubes in the foil until no light can enter. Set these in a safe place where they will not be bumped or moved for the experiment.

6. Select one test tube containing the elodea plant and select a test tube with no plant. Set these test tubes on a window sill or area where they will receive full direct sunlight for the experiment.

7. Set your last test tubes (one containing the elodea plant and one without) in an area that gets indirect light.

8. Record your observations at 24, 48, and 72 hours.

Name _____

Questions

1. Explain what happened in this lab, using complete sentences. Be sure to provide supporting data or statements.

   _____
   _____
   _____
   _____
   _____
   _____

2. Do plants consume or release more carbon dioxide?

   _____
   _____

Name _____ Date _____ Class _____

LESSON 8.6 Reproductive Parts of a Plant

Lesson Review

Carefully study the lesson and then answer the following questions.

1. List the four main parts of a flower.

    A. _____

    B. _____

    C. _____

    D. _____

2. How do the sepals protect the flower as it develops? _____
_____
_____
_____

3. What purpose do flower petals serve? _____
_____
_____

4. List the two main parts of the stamen.

    A. _____

    B. _____

5. Why is the anther an important part of the flower? _____
_____
_____

6. List the three main parts of the pistil.

    A. _____

    B. _____

    C. _____

7. How are flowers pollinated? _____
_____
_____
_____

8. When the pollen of one plant pollinates the flower of another plant, this is called _____.
_____

Name _____

9. What is the difference between a complete flower and an incomplete flower? _____
_____
_____
_____

10. What is the difference between a perfect flower and an imperfect flower? _____
_____
_____
_____

11. List three examples of imperfect flowers.

   A. _____

   B. _____

   C. _____

 Activity 8.6A

Finding Flowers

Perfect and imperfect, complete and incomplete–flowers are all around you! Have you ever stopped to look at them? Using the knowledge you gained about flowers from Lesson 8.6, you are going to see what sorts of flowers you can find.

Take a walk around your school grounds, home, or neighborhood, stopping to look at the plants you pass. Where are their flowers? Looking closely, take photographs or make sketches of the flowers that you find. Once you have a handful of flowers photographed or sketched, stop to identify the flower parts.

Did you find more complete or incomplete flowers? Did you find perfect or imperfect flowers? Do some kinds appear to be more common than others? What conclusions can you draw from your observations?

Name _____ Date _____ Class _____

LESSON 8.7
Sexual Propagation

Lesson Review

Carefully study the lesson and then answer the following questions.

1. Why is growing plants from seed not always the best method of propagation? _____
   _____
   _____

2. Why is growing plants from seed often desirable? _____
   _____
   _____

3. How do the germination requirements for seeds vary? _____
   _____
   _____

4. _____ Many seeds with a hard seed coat must go through the process of _____ prior to germination.
   A. dormancy
   B. stratification
   C. scarification
   D. soaking

5. Do you think that plants requiring stratification for reproduction are more common in warm or cold areas? Why do you think so? _____
   _____
   _____

6. List four factors that should be considered prior to direct seeding.

   A. _____
   B. _____
   C. _____
   D. _____

7. _____ _____ seeding occurs when a machine or person places seeds in the ground where they will continue to grow.
   A. Direct
   B. Indirect
   C. Funky fresh
   D. Permanent

8. What is one main advantage of indirect seeding? _____
   _____
   _____

Copyright Goodheart-Willcox Co., Inc.
May not be reproduced or posted to a publicly accessible website.

9. When does it make sense to transplant seedlings from their starter containers? _____
_____
_____

10. What information might you find on a seed packet? _____
_____
_____

Name _____ Date _____ Class _____

 **Activity 8.7A**

Seed Viability Lab

Have you ever planted a seed, and then ... nothing? Seeds may be a cheap, easy, or convenient way to grow plants, but not every seed sprouts. Why not? Seeds need to be stored properly to maintain their viability. Seeds stored at room temperature will last for a season; prolonged storage requires lower temperatures. Humidity also affects seed viability.

Supplies
- Bean seeds (Any kind will work, but large beans are easier to see)
- TTC (triphenyltetrazolium chloride)
- Gloves
- Beaker
- Paper towels or napkins
- Razor blade

Instructions
1. Place the bean seeds in a beaker. Cover them with TTC and let them soak overnight.
2. Put on your safety gloves!
3. Remove the bean seeds from the beaker and place them gently into your paper towels or napkins.
4. Using the razor blade carefully and pointing it away from you, cut open the seeds.

Questions
1. What do you see when you cut open the seeds? Are all of the seeds the same?

   _____
   _____
   _____

2. Seeds with living tissue should be red after being soaked in the TTC. How does that information help you interpret what you see?

   _____
   _____
   _____

3. Where did you get your bean seeds? How do you think your results would differ if you got them from the dried goods section of the grocery store? From a seed packet?

   _____
   _____
   _____
   _____

Name _____ Date _____ Class _____

LESSON 8.8 Asexual Propagation

Lesson Review

Carefully study the lesson and then answer the following questions.

1. Name three plants that do not produce flowers.

   A. _____

   B. _____

   C. _____

2. What are the main benefits of using asexual propagation? _____
   _____
   _____

3. _____ What is the most commonly used method of asexual propagation in the horticulture industry?
   A. Cuttings
   B. Separation
   C. Division
   D. Grafting

4. What conditions must be considered when taking cuttings? _____
   _____
   _____

5. Stem cuttings can be classified as _____ or _____.
   _____

6. What types of plants are most commonly produced via leaf cuttings? _____
   _____

7. _____ and _____ are often successfully propagated through separation.
   _____
   _____

8. _____ and _____ are often successfully propagated through division.
   _____
   _____

9. _____ In grafting, the portion with the stem and bud(s) is called the:
   A. rootstock
   B. root
   C. scion
   D. growth node

10. What is required for grafting to be successful? _____
    _____

Name _____ Date _____ Class _____

 # Activity 8.8A

Understanding Separation and Division

It may seem difficult to tell when separation and division will work to propagate a plant, so perhaps doing some research and looking at visual aids will help!

Using a device with internet access, search for images of corms and bulbs. (You may need to search for "plant bulbs" to avoid being inundated with images of light bulbs!) You learned in this lesson that plants grown from corms and bulbs are good candidates for separation. Why do you think this is?

1. List five plants grown from corms.

   _____
   _____
   _____

2. List five plants grown from bulbs.

   _____
   _____
   _____

Going back to your device, search for images of rhizomes and tubers. You learned in this lesson that plants grown from rhizomes and tubers are good candidates for division. Why do you think this is?

1. List five plants grown from rhizomes.

   _____
   _____
   _____

2. Live five plants grown from tubers.

   _____
   _____
   _____

3. If you saw an unknown plant with a modified rootstock of some type, like those listed on this page, do you think you would be able to tell how to propagate it? How?

   _____
   _____
   _____

Name _____ Date _____ Class _____

LESSON 8.9 Agricultural Pests

Lesson Review

Carefully study the lesson and then answer the following questions.

1. How do pests cause issues for plants? _____
   _____
   _____
   _____

2. What defines an insect? _____
   _____
   _____

3. _____ Approximately what percentage of insect species found in the United States cause damage to agricultural crops?
   A. 10 percent
   B. 2 percent
   C. 1 percent
   D. One-tenth of 1 percent

4. _____ What determines the type of damage that insects will cause to plants?
   A. The number of legs
   B. The type of mouthparts
   C. How quickly they move
   D. The size of the insect

5. Are weeds inherently noxious or bad? Explain your answer. _____
   _____
   _____
   _____

6. Why are many weeds successful at persisting in landscapes, despite the fact that people don't want them?
   _____
   _____

7. What are the three conditions necessary for an abiotic plant disease to occur?
   A. _____
   B. _____
   C. _____

8. _____ What type of pathogen causes the most plant diseases?
   A. Fungi
   B. Insects
   C. Bacteria
   D. Viruses

Name _____

9. List three plant diseases caused by fungi.

   A. _____

   B. _____

   C. _____

10. List three plant diseases caused by bacteria.

    A. _____

    B. _____

    C. _____

11. List three plant diseases caused by viruses.

    A. _____

    B. _____

    C. _____

12. Integrated pest management includes which of these? Check all that apply.

    A. _____ Biological controls

    B. _____ Chemical controls

    C. _____ Cultural controls

    D. _____ Mechanical controls

    E. _____ Environmental controls

Name _____ Date _____ Class _____

 Activity 8.9A

Integrated Pest Management Activity

It's one thing to read about managing pests; it's another thing entirely to put a plan in place to do so. What all is required? Well, you're about to find out!

For this activity, you'll be assigned a team. Each team will be assigned two insects to consider. (Group 1: aphids and fungus gnats; Group 2: leaf minor and mealy bugs; Group 3: scale and slugs; Group 4: thrip and whiteflies; Group 5: spider mites and leaf hoppers)

As a group, you are going to create an integrated pest management plan! Your plan should include:
- Description of the pests you were assigned, including their mouthparts
- Description of the damage these pests can cause to crops
- Description of the pests' life cycles
- Biological controls that can be taken to address this issue, including any potential problems or issues
- Chemical controls that can be taken to address this issue, including any potential problems or issues
- Cultural controls that can be taken to address this issue, including any potential problems or issues
- Mechanical controls that can be taken to address this issue, including any potential problems or issues
- Cost for implementing this proposal

Have some fun with it! When in doubt, more details are always better, so add information where you can. Be prepared to present your plan to the class!

Name _____ Date _____ Class _____

LESSON 9.1 Fundamentals of Animal Science

Lesson Review

Carefully study the lesson and then answer the following questions.

1. What is animal science, and what is the difference between livestock and companion animals? _____
   _____
   _____

2. _____ What do you call an animal that only consumes plant-based material?
   A. Herbivore
   B. Carnivore
   C. Nomadic
   D. Fodder

3. _____ What do you call an animal that only eats meat?
   A. Herbivore
   B. Carnivore
   C. Nomadic
   D. Fodder

4. What are genetically determined characteristics that are specific and desirable? _____
   _____
   _____
   _____

5. _____ What word describes the lifestyle of livestock caretakers who moved from place to place to find food?
   A. Herbivore
   B. Carnivore
   C. Nomadic
   D. Fodder

6. Explain what temperament is and why it is important in animal agriculture. _____
   _____
   _____

7. _____ What is the food that is given (as opposed to foraged for by) to domesticated animals such as cattle, sheep, or horses?
   A. Herbivore
   B. Carnivore
   C. Nomadic
   D. Fodder

Copyright Goodheart-Willcox Co., Inc.
May not be reproduced or posted to a publicly accessible website.

8. Which of the following animals are carnivores? Check all that apply.

   A. _____ Cows

   B. _____ Dogs

   C. _____ Cats

   D. _____ Sheep

9. What percentage of protein that humans consume comes from animal sources? _____
   _____

10. _____ Which of the following animals are herbivores? Circle all that apply.
    A. Cows
    B. Dogs
    C. Cats
    D. Sheep

11. Give a brief overview of the history and importance of livestock production and the role humans have played in its progress over time. _____
    _____
    _____

12. Summarize factors that make some animals more suitable for domestication than others. _____
    _____
    _____

13. Explain factors that one should consider if they are interested in raising livestock. What are the basic requirements and things they should keep in mind? _____
    _____
    _____
    _____

Name _____ Date _____ Class _____

Activity 9.1A

The History of Livestock

The domestication and use of livestock as a source of protein for humans goes back as far as human history does. Pick one species of livestock that interests you, and research the history, uses, needs, and other interesting facts about your chosen species to present to your classmates!

Name _____ Date _____ Class _____

LESSON 9.2 The Importance of the Livestock Industry

Lesson Review

Carefully study the lesson and then answer the following questions.

1. Who were the first to bring beef cattle to North America in large numbers, and in what year did they do so?
_____

2. Who were the first to bring dairy cattle to North America, and in what year did they do so? _____
_____

3. Describe the two main differences between *Bos taurus* and *Bos indicus* cattle. _____
_____

Match the breeds with the correct scientific name.

4. _____ Holstein   A. *Bos indicus*
5. _____ Santa Gertrudis B. *Bos taurus*
6. _____ Angus
7. _____ Hereford
8. _____ Brahman

9. _____ Where were the first dairy cattle in the United States introduced?
   A. Delaware
   B. Massachusetts
   C. Virginia
   D. Maryland

10. _____ Where were the first beef cattle in the United States introduced?
    A. Delaware
    B. Massachusetts
    C. Virginia
    D. Maryland

11. _____ What type of cattle are known for milk production and typically of the species *Bos taurus*?
    A. Beef
    B. Sheep
    C. Hereford
    D. Dairy

12. _____ What are the annual sales of the US dairy industry?
    A. $400 million
    B. $600 billion
    C. $500 billion
    D. $500 million

13. According to scientists, around how many years ago were horses first domesticated? _____
_____

118 Exploring Agriculture Food and Natural Resources Lab Workbook

Name _____

14. In your own words, describe the term *finishing*. _____
_____
_____

15. _____ How many different identified vocalizations do pigs make?
    A. Only 1
    B. Less than 10
    C. More than 20
    D. At least 30

16. _____ Which livestock sector most recently evolved into commercial production?
    A. Beef
    B. Poultry
    C. Dairy
    D. Swine

17. Will goats really eat anything? _____
_____

18. _____ Which livestock species is believed to be one of the first to be domesticated?
    A. Horses
    B. Sheep
    C. Fowl
    D. Cattle

 Activity 9.2A

Livestock Species, Large and Small

This lesson unveiled the different types of livestock species and gave us a brief introduction and background to each. Let's take what we've learned so far a step further! Grab a partner and select a species of large or small livestock to research. Make a poster or digital presentation to present to your classmates that includes the following information:

- Where your species was domesticated
- When it was first introduced in the United States
- What it is known for and the products it produces
- Where it lives in the United States today
- Scientific name
- Three breed examples
- Lots of visuals

Name _____ Date _____ Class _____

LESSON 9.3 Major Breeds of Livestock

Lesson Review

Carefully study the lesson and then answer the following questions.

1. What is a breed? Name three example characteristics that can help sort an animal into a breed. _____
   _____
   _____
   _____

2. _____ Which of the following is not considered a type of poultry?
   A. Chicken
   B. Hawk
   C. Duck
   D. Goose

3. _____ What unit of measurement is used for horses' height?
   A. Marble
   B. Poll
   C. Horn
   D. Hand

4. _____ What is the term for a desirable mixture of fat and lean meat?
   A. Marbled
   B. Polled
   C. Horned
   D. Hand

5. _____ Which term means "without horns"?
   A. Marbled
   B. Polled
   C. Horned
   D. Hand

6. _____ What is the term for meat that comes from sheep?
   A. Beef
   B. Boer
   C. Mutton
   D. Duroc

Copyright Goodheart-Willcox Co., Inc.
May not be reproduced or posted to a publicly accessible website.

Match the breed description with the correct breed name from the word bank below.

7. _____ A red breed of pigs with forward-drooping ears that are known for fast growth and quality muscling.

8. _____ A common dairy goat breed originally from Africa with the highest milk fat content of any dairy goat breed.

9. _____ Believed to be one of the oldest dairy cow breeds, these animals are a very large breed, solid brown, with high protein content in their milk.

10. _____ Ancient middle eastern horse breed known for their high tail set, wide-set eyes, broad forehead, and arched neck among other prominent characteristics.

11. _____ Originating from Scotland, this beef breed is the most popular breed in the US that is known for being well-marbled.

12. _____ A meat goat breed originating from South Africa that are white with a red head and white blaze down their face.

13. _____ Hogs that originated from York County, England that are solid white in color and known as the "mother breed" due to litter size and excellent mothering ability.

14. _____ This breed of cattle originated from Switzerland, and is known for producing meat and milk. It can vary in color.

15. _____ A breed of sheep that are white with black heads and legs that are known for growth and muscling.

16. _____ This horse breed is known for their ability to carry weight at a sustained speed over long distances.

17. _____ This is the most popular and well-recognized of the dairy breeds and are quite large with black and white or even red and white coloring, as well as the highest milk production per cow.

18. _____ A solid white breed of sheep that are heavily muscled and known for being dual purpose as high producers of mutton and wool alike.

A. Holstein
B. Brown Swiss
C. Black Angus
D. Suffolk
E. Simmental
F. Yorkshire
G. Nubian
H. Boer
I. Thoroughbred
J. Duroc
K. Arabian
L. Dorset

 Activity 9.3A

Poultry Breeds Memory Game

We've tested our knowledge of breeds of various species, and now let's focus on poultry breeds! Taking index cards, write the name of each of the different poultry breeds discussed in this lesson (chickens, ducks, geese, and turkeys). On the other side of each index card, write a unique and identifiable characteristic for that breed, or even attach a photo or picture. Once your cards are made, they can be used individually to hone your breed knowledge and recognition skills, or you can partner up to practice with a friend! If you want to create a longer or more complicated deck, you can build multiple cards per breed, or expand your deck to include other types of livestock species.

Name _____ Date _____ Class _____

LESSON 9.4
Introduction to Livestock Selection

Lesson Review

Carefully study the lesson and then answer the following questions.

1. Describe the difference between a breeding animal and a market animal. _____
   _____
   _____
   _____
   _____

2. _____ What is the milk-producing organ of cattle?
   A. Dam
   B. Cull
   C. Udder
   D. Conformation

3. The male parent is also called the _____.
   _____

4. The female parent is also called the _____.
   _____

5. What are chickens raised to produce table eggs called? _____
   _____

6. _____ are a method of collecting production data on livestock to select the most productive animals.
   _____

7. _____ To _____ is to remove an animal from a herd.
   A. dam
   B. cull
   C. udder
   D. conform

8. What are the four basic characteristics that are evaluated in all species?

   A. _____   C. _____

   B. _____   D. _____

9. _____ Quality muscle shape or structure is evaluated as:
   A. A sire
   B. A market animal
   C. Culling
   D. Conformation

124 Exploring Agriculture Food and Natural Resources Lab Workbook

Name _____

10. What five factors are considered for horse selection?

    A. _____ D. _____

    B. _____ E. _____

    C. _____

11. What are chickens raised for meat production called? _____
    _____

12. What are six data points or pieces of information recorded in livestock performance records?

    A. _____ D. _____

    B. _____ E. _____

    C. _____ F. _____

# Activity 9.4A

Making Comparisons

While the general concept of selection is the same across all species of livestock, each species differs quite a bit when it comes to the specifics. Pick a species of livestock and create a chart that displays the following information to be presented to your classmates:

- Species
- What they are raised for (meat, fiber, milk, eggs, dual purpose, etc.)
- Selection characteristics used, and what characteristics are most important
- Type of scale, if any, used to determine if an animal is good, bad, or mediocre for that selection characteristic
- Visually identifiable characteristics that may exist

Name _____ Date _____ Class _____

Livestock Products and By-Products

Lesson Review

Carefully study the lesson and then answer the following questions.

1. _____ What is the process of using heat to sterilize milk so it is safer for consumption?
   A. Conformation
   B. Pasteurization
   C. Offal
   D. Collagen

2. What is the difference between *fertilized* and *unfertilized* eggs, and how are each used? _____
   _____
   _____
   _____

3. _____ What country is the top wool-producing country in the world, and what percent of all the wool in the world do they produce annually?
   A. United States, 25%
   B. Australia, 20%
   C. India, 25%
   D. Australia, 25%

4. _____ What is another name for the whole organs, like the liver, that butcher shops sell?
   A. Variety meats
   B. Abbatoir
   C. Collagen
   D. Carcass

5. What is offal and why is it an important part of the animal? _____
   _____
   _____
   _____

6. What are the two classifications of livestock by-products and what do they mean? _____
   _____
   _____
   _____

7. _____ What is edible by-product is the structural protein found in connective tissues, like bone and skin?
   A. Variety meats
   B. Abbatoir
   C. Collagen
   D. Carcass

Copyright Goodheart-Willcox Co., Inc.
May not be reproduced or posted to a publicly accessible website.

8. _____ What is another general name for secondary animal products?
    A. Variety meats
    B. Collagen
    C. By-products
    D. Carcass

9. What are the two most consumed meats worldwide? _____
_____

10. _____ What country produces more than a fifth of the world's milk?
    A. United States
    B. China
    C. India
    D. France

11. What is the term for an animal processing facility? _____

Name _____ Date _____ Class _____

 **Activity 9.5A**

Making Connections: Animal By-Products in Our Everyday Lives

Animal by-products are much more than our favorite cheeses, eggs for breakfast, or our favorite leather shoes. By-products are found in an incredible amount of our everyday products, most of which you'd probably never even consider, like medicine, makeup, and Jell-O! Use the internet to find four surprising items or products that contain animal by-products and present your most interesting finding(s) to the class, including what animal the by-product came from.

Name _____ Date _____ Class _____

LESSON 9.6 Livestock Physiology

Lesson Review

Carefully study the lesson and then answer the following questions.

1. Where is cardiac muscle found in the body? _____
   _____

2. Explain the difference between monogastric and ruminant digestive systems. _____
   _____
   _____

3. _____ What system are skeletal muscles part of?
   A. Skeletal system
   B. Muscular system
   C. Nervous system
   D. Circulatory system

4. _____ Which is the term used for the tissue that attaches muscle to bone?
   A. Cartilage
   B. Tendon
   C. Marrow
   D. Ligament

5. _____ What is the branch of biology that deals with the normal functions of living organisms and their parts?
   A. Anatomy
   B. Physiology
   C. Ruminant
   D. Monogastric

6. _____ Which term refers to the soft center of the bone?
   A. Cartilage
   B. Tendon
   C. Marrow
   D. Ligament

7. _____ Which is the term used for the tissue that attaches bone to bone?
   A. Cartilage
   B. Tendon
   C. Marrow
   D. Ligament

8. What is another popular way of referring to rumination? _____
   _____

9. _____ Which term refers to a tough and flexible connective tissue that provides support between the bones at joints?
   A. Cartilage
   B. Tendon
   C. Marrow
   D. Ligament

Name _____

Carefully study the lesson and then match the system with its function.

10. _____ The _____ is responsible for transferring information throughout the body.

11. _____ The _____ controls a body's movement.

12. _____ The _____ manages oxygen intake and absorption.

13. _____ The _____ moves blood and nutrients throughout the body.

14. _____ The _____ provides structure, protection, and support for the body.

A. Skeletal system
B. Muscular system
C. Respiratory system
D. Circulatory system
E. Nervous system

 Activity 9.6A

Digesting Livestock Physiology

It's time to regurgitate what we've learned in Lesson 9.6. As a class, split into three groups. Each group will cover one of the three types of digestive systems found in livestock (avian, ruminant, and monogastric). Using the textbook and internet resources to conduct research, create a poster with a drawing of the assigned digestive system. Be sure to include species examples in your poster. Present your findings to the other two groups in your class.

Name _____ Date _____ Class _____

Livestock Feeding and Nutrition

Lesson Review

Carefully study the lesson and then answer the following questions.

1. What are the six major nutrients needed by livestock?

   A. _____   D. _____

   B. _____   E. _____

   C. _____   F. _____

2. How do *roughages* and *concentrates* differ? _____
   _____
   _____

3. Why should animal feed be *palatable*? _____
   _____
   _____

4. What does GRAS stand for? _____
   _____

Carefully study the lesson and then match the letter of the term with its correct definition.

5. _____ This nutrient makes up 60–70 percent of the body of most livestock, and without it, livestock would perish within days.

6. _____ These are sugars and starches and are easily digested.

7. _____ These are made up of carbon, hydrogen, oxygen, and nitrogen, and serve as the building blocks of protein.

8. _____ This is one of the main categories of roughages.

9. _____ This is another term for fats and oils and provides energy to produce body heat. They are made up of carbon, hydrogen, and oxygen.

10. _____ Substances that provide nourishment for growth and the maintenance of life.

11. _____ These are the main energy nutrient found in livestock rations.

12. _____ Compounds made up of amino acids.

13. _____ Also called fiber, these are made up of cellulose and lignin, making them more difficult to digest than other nutrients.

A. Proteins
B. Water
C. Simple carbohydrates
D. Nutrients
E. Amino acids
F. Carbohydrates
G. Complex carbohydrates
H. Lipids
I. Legumes

Copyright Goodheart-Willcox Co., Inc.
May not be reproduced or posted to a publicly accessible website.

Name _____ Date _____ Class _____

 Activity 9.7A

Livestock Feed Labels in Action

Pick your favorite species of livestock we've looked at thus far. What is their purpose? After reviewing Lesson 9.7, what nutrients do you think are most important to your selected species' growth? Using the internet, research feed for your chosen animal and compare nutrient contents between the products of multiple competitors. Which one would you select and why? What are the differences, if any?

Name _____ Date _____ Class _____

Keeping Livestock Healthy

Lesson Review

Carefully study the lesson and then answer the following questions.

1. _____ What term refers to disease(s) easily transmitted from one animal to another?
   A. Infectious
   B. Contagious
   C. Zoonotic
   D. Parasites

2. _____ What term refers to organisms that live in, with, or on another organism (host)?
   A. Infectious
   B. Contagious
   C. Zoonotic
   D. Parasites

3. What is an effective means of primary prevention that stimulates an animal's immune system and protects from future infection? _____
   _____

4. _____ What term refers to disease(s) that can be transferred between humans and animals?
   A. Infectious
   B. Contagious
   C. Zoonotic
   D. Parasites

5. What is another term for isolating sick or new animals to keep other animals from getting sick? _____
   _____

6. What are the three types of prevention?

   A. _____

   B. _____

   C. _____

7. _____ What term refers to disease(s) that can infect or cause harm to another creature?
   A. Infectious
   B. Contagious
   C. Zoonotic Disease
   D. Parasites

8. What is a *disease*? _____
   _____

9. List five characteristics of good livestock health management practices:

   A. _____   D. _____

   B. _____   E. _____

   C. _____

Copyright Goodheart-Willcox Co., Inc.
May not be reproduced or posted to a publicly accessible website.

10. How does health play a role in the livestock industry and food production? _____
_____
_____

11. What are the three main types of diseases in livestock?

   A. _____

   B. _____

   C. _____

Activity 9.8A

Disease Discovery

To explore some of the worrisome and downright dangerous diseases that farmers must be concerned with in their herds, let's begin by selecting one of the three main types of diseases (infectious, parasitic, or those caused by improper nutrition). Then, go online to select a disease from the category you selected. Dive deeper by looking for answers to the following questions:

- Which of the three main types of disease categories does your disease fit into?
- What species does it affect? There may be more than one.
- Is it considered to be a zoonotic disease?
- What symptoms accompany the disease?
- How can this disease be treated? Does it require veterinary care?
- How much money can this disease cost an individual farmer or the industry as a whole?
- What are some ways that farmers can protect and prevent against this disease?

Once you've found these, make a digital presentation and deliver it to the class!

Name _____ Date _____ Class _____

Safety around Livestock

Lesson Review

Carefully study the lesson and then answer the following questions.

1. _____ What part of the US Department of Labor keeps statistics on workplace safety and works to make all workplaces as safe as they can be?
   A. Department of Agriculture
   B. Occupational Safety and Health Administration
   C. Farm Service Agency
   D. Department of Workplace Safety

2. _____ What is the term for the way you present your emotions in your actions?
   A. Experience
   B. Demeanor
   C. Flight Zone
   D. PPE

3. By behaving _____ and _____, you can help to minimize the risk of accidents for you and the animals you work with.
   _____
   _____

4. _____ What is a halter?
   A. A muzzle that covers the animal's mouth
   B. A collar that goes around the animal's neck
   C. A type of collar that fits behind the ears and around the muzzle
   D. A leash that controls the animal

5. What does PPE stand for and what is its purpose? _____
   _____
   _____

6. What is livestock confinement housing? _____
   _____
   _____

7. What is a flight zone? _____
   _____
   _____

8. _____ What do you call practical contact with and observation of facts or events?
   A. Experience
   B. Demeanor
   C. Flight zone
   D. PPE

Name _____

9. What hazards may livestock pose to human workers, and how can these hazards be mitigated? _____
_____
_____

10. List five examples of PPE:

    A. _____   D. _____

    B. _____   E. _____

    C. _____

11. Identify and describe three safety hazards that are associated with livestock facilities. _____
_____
_____
_____

Name _____ Date _____ Class _____

 Activity 9.9A

Safety First, Always!

Partner up! We're going to look into the details of seven major safety concerns when working with livestock. To begin, you and your partner will select one of the following categories: fire safety, grain storage, chemical use, animal demeanor, mechanical equipment, air quality, and sanitation. Using the text and internet, research the major safety concerns within your designated area. Prepare a presentation, with visual aids, on the hazards in your category, and best practices for safety. To ensure your classmates will be properly informed after your presentation, be sure to include ways they can mitigate the hazards you identify!

Name _____ Date _____ Class _____

LESSON 9.10 Specialty Livestock Production

Lesson Review

Carefully study the lesson and then answer the following questions.

1. How many stomach chambers do pseudo-ruminants have? _____

2. What makes bison an attractive alternative to beef? _____

3. Llamas are _____ whose hair does not contain lanolin. This makes their hair _____, or relatively unlikely to cause an allergic reaction.

4. List four examples of specialty animals and their products. _____

5. _____ Which specialty livestock species' meat has roughly 18 percent higher protein content than chicken?
   A. Quail
   B. Rabbit
   C. Bison
   D. Fish

6. Explain what it means to be classified as specialty livestock. _____

7. _____ is the farming of snails for human consumption and cosmetic production.

8. _____ are even-toed mammals with a three-compartment stomach.

9. _____ is the production or culture of aquatic organisms under controlled conditions.

10. _____ is the growth and harvesting of saltwater species.

11. _____ is the care and management of honey bees to produce honey and wax.

Copyright Goodheart-Willcox Co., Inc.
May not be reproduced or posted to a publicly accessible website.

Activity 9.10A

Apiculture, Aquaculture, Agriculture, Oh My!

Whoa, the realm of animal agriculture just got a whole lot bigger, huh? Livestock are much more than just cows and sows, and although these specialties can be small, they can be mighty. Reflecting back to the beginning of this text, we discovered that our world's population is rapidly increasing, and we'll need to increase our food supply by over half in order to survive. What are some ways these niche specialty livestock species can help with that growing dilemma? Review the chapter once more and choose a species to represent in a friendly debate with a classmate. Use the internet and other resources to prepare your argument for how it may be utilized. Don't forget to consider the potential issues it might pose. Have a debate with a classmate about whether that species could be scaled up to become a larger part of agriculture. Take turns exchanging ideas.

Name _____ Date _____ Class _____

LESSON 9.11 / Companion Animals

Lesson Review

Carefully study the lesson and then answer the following questions.

1. What is the role of the companion animal industry in agriculture? _____
   _____
   _____

2. _____ Companion animals are a _____ industry in the United States.
   A. $2 billion
   B. $3 billion
   C. $4 million
   D. $5 billion

3. What are four examples of companion animals? _____
   _____

4. _____ What were cats likely originally domesticated for?
   A. Rodent control
   B. Companionship
   C. Good luck
   D. Religious purposes

5. _____ What type of animal was first tamed for companionship?
   A. Cats
   B. Horses
   C. Dogs
   D. Fish

6. _____ *True or False?* It can be illegal to own a wild animal as a pet.

7. Describe factors that should be taken into consideration when choosing an appropriate companion animal.
   _____
   _____
   _____

8. What is a service animal? _____
   _____

9. What factors should always be considered to be a responsible pet owner? _____
   _____
   _____

10. Service animals are almost exclusively _____.
    _____

11. Why are companion animals raised? _____
    _____

Copyright Goodheart-Willcox Co., Inc.
May not be reproduced or posted to a publicly accessible website.

Name _____ Date _____ Class _____

Activity 9.11A

The Humble Beginnings of Humankind's Favorite Friends

When we think of "man's best friend," we often think of dogs and their incredible loyalty, or fuzzy felines curled up with you on the couch, or maybe even a chatty parakeet that loves to hang out on your shoulder. Were these animals always so friendly? How did these animals get to such a familiar and friendly place with people? Pick your favorite companion animal and research their history. What was the species originally domesticated for, if not for companionship? Be sure to incorporate historical images to show the progress and evolution of their relationships with humans over time.

Name _____ Date _____ Class _____

LESSON 10.1 Wildlife and Natural Resources Careers

Lesson Review

Carefully study the lesson and then answer the following questions.

1. A(n) _____ is an area where the impact of human activities is minimal or nonexistent.
   _____

2. _____ What government department handles the task of managing the nation's wild areas?
   A. Department of the Interior
   B. National Park Service
   C. Department of Natural Resources
   D. Department of Wildlife, Forestry, and Natural Resources

3. _____ Which area involves managing, conserving, and creating forests?
   A. Natural Resources
   B. Fish and Wildlife Management
   C. Forestry
   D. Dendrology

4. A(n) _____ is a protected place for wildlife.
   _____

5. List four examples of careers within WFNR.
   A. _____
   B. _____
   C. _____
   D. _____

6. What knowledge, skills, and dispositions are suggested for those interested in careers within WFNR?
   _____
   _____
   _____

Carefully study the lesson and then answer the following questions by matching the appropriate letter with its definition.

7. _____ The study of rocks that make up the earth.

8. _____ The study of botany of woody plants.

9. _____ The study of the movement of water on Earth's surface.

10. _____ The study of Earth's weather patterns.

11. _____ The creation of maps, using global positioning systems.

12. _____ Bringing nature and people closer together, into a relationship where both can benefit—i.e., city planners, city park rangers, horticulturalists, etc.

13. _____ Looks for ways to conserve and protect natural ecosystems.

A. Environmental management
B. Cartography
C. Hydrology
D. Geology
E. Dendrology
F. Meteorology
G. Ecological management

Name _____ Date _____ Class _____

 Activity 10.1A

Urban Environmental Management

Environmental management isn't a task only for those that work in protected wild areas like state and national parks. Environmental management is for everyone in every place, even in urban areas! Let's imagine you work in the field of environmental management and have just been hired by your city to design a new greenspace area. Using posterboard or technology along with your knowledge of your city, design a greenspace that incorporates natural elements like water, grassy areas, trees, shrubbery, and flowers. Keep in mind that it needs to be accessible, usable, and environmentally friendly. For example, it could be beneficial to create a pollinator garden that has informational signs throughout. What type of trees might you consider using? Will you include paths? Get creative! Who knows, maybe your design will inspire your city to make your vision a reality! Share your design with the class.

Name _____ Date _____ Class _____

LESSON 10.2 Understanding and Researching WFNR Issues

Lesson Review

Carefully study the lesson and then answer the following questions.

1. _____ is an unreasoned outlook or opinion about something.
   _____

2. _____ are objective, meaning they are widely known to be accurate and true.
   _____

3. _____ is communication that attempts to use your emotions and feelings to lead you to think a certain way about something.
   _____

4. _____ is a term describing how Earth is becoming gradually warmer due to human activity.
   _____

5. _____ is defined as electronic communication between members of an online social network created to share information, ideas, messages, and other content. It can be the source of a lot of misinformation.
   _____

6. _____ are often based on emotions and feelings.
   _____

7. A(n) _____ is a problem that typically involves a dispute between two or more parties and addresses topics of concern to a broad range of people.
   _____

8. List and describe two important research skills:
   A. _____
   B. _____

9. What are three ways to guard against personal bias?
   A. _____
   B. _____
   C. _____

10. Explain how propaganda and opinions are connected to one another. _____
    _____
    _____
    _____

Copyright Goodheart-Willcox Co., Inc.
May not be reproduced or posted to a publicly accessible website.

Name _____ Date _____ Class _____

Activity 10.2A

Real Examples of Propaganda, Issues, and Bias

Let's take what we've learned in Lesson 10.2 and put it into action! Propaganda, issues, and bias are found everywhere, not just in WFNR. Search the internet to find an example of propaganda, issues, and/or bias. Identify where those are demonstrated within the article, social media post, or advertisement. What might have been a more accurate way to share the information? Share your findings with your classmates.

Name _____ Date _____ Class _____

LESSON 10.3 Ecosystems

Lesson Review

Carefully study the lesson and then answer the following questions.

1. What is biodiversity and why is it essential? _____
   _____
   _____

2. _____ What are the first plants to inhabit an old field?
   A. Aquifers
   B. Ecosystems
   C. Pioneer species
   D. Succession

3. _____ is water that soaks into the ground.
   _____

4. The _____ is the natural process of converting living organisms into organic and inorganic matter.
   _____

5. _____ is the process of water molecules changing from liquid to gas.
   _____

6. _____ What do we call concentrated quantities of groundwater deep beneath the Earth's surface?
   A. Aquifers
   B. Ecosystems
   C. Pioneer species
   D. Succession

7. _____ What is a community of living organisms that have a beneficial relationship?
   A. Aquifers
   B. Ecosystems
   C. Pioneer species
   D. Succession

8. _____ eat plant and animal matter.
   _____

9. _____ What is the term for the gradual and long-term changes that ecosystems are constantly going through?
   A. Aquifers
   B. Ecosystems
   C. Pioneer Species
   D. Succession

10. _____ is the process of changing water vapor into a liquid.
    _____

11. _____ happens when dead plants and animals break down into simpler matter such as water, sugars, and minerals.
    _____

Copyright Goodheart-Willcox Co., Inc.
May not be reproduced or posted to a publicly accessible website.

Name _____ Date _____ Class _____

Activity 10.3A

The Water Cycle

We learned in Lesson 10.3 that the water cycle is essential to the survival of all living things. It's important to know the process from start to finish and truly understand each step! To help us do so, let's create a flow chart that demonstrates the steps of the water cycle. You can do this by designing a simple flow chart that shows a step-by-step process, or by using graphics like Figure 10.3.7 in your text. Make sure you include descriptions for each labeled step. This activity can be done using posters or website technology, like an interactive Prezi presentation to show to your classmates.

Name _____ Date _____ Class _____

Creating Conservation Plans

Lesson Review

Carefully study the lesson and then answer the following questions.

1. _____ are bodies of water where freshwater from inland sources meets saltwater from the ocean and are notoriously sensitive habitats.
   _____

2. _____ tell us about the environmental conditions in which they live, using an increase or decline in their population levels as a metric.
   _____

3. _____ are the intermediate steps for achieving goals.
   _____

4. _____ is putting time, money, and resources into proactively cleaning and fixing an area.
   _____

5. _____ are put in place to protect ecosystems or species within those ecosystems.
   _____

6. _____ is the careful use of resources to prevent unnecessary waste.
   _____

7. What does SMART stand for?
   A. S _____
   B. M _____
   C. A _____
   D. R _____
   E. T _____

8. Explain the purpose of conservation plans. _____

9. What are the steps to developing a conservation plan?
   A. _____
   B. _____
   C. _____
   D. _____
   E. _____
   F. _____
   G. _____
   H. _____

Putting a Conservation Plan in Action

According to the International Union of Conservation and Nature, in 2021 there were believed to be 16,300 endangered species and at least 38,500 species under threat. Select one of these species that interests you and develop a conservation plan to help them. If there is already one in place, you can either create a new one or critique the existing plan to incorporate new ideas. Make a presentation to present your new conservation plan or the conservation plan in place to your fellow classmates. Be sure to include information about your species that elicits emotion that will make your classmates care about this initiative. Consider the impact this species has on the environment as well!

Name _____ Date _____ Class _____

LESSON 10.5 The Wildlife, Forestry, and Natural Resources Industry

Lesson Review

Carefully study the lesson and then answer the following questions.

1. _____ What is the term used for a practice that acts on a wildlife species by increasing or decreasing its population through changes to the food supply, changing the habitat, or managing predators?
   A. Carrying capacity
   B. Silviculture
   C. Manipulative management
   D. Custodial management

2. What is nuclear energy and how is it unique? _____
   _____
   _____

3. _____ What is the term for a type of forest management that controls the growth, health, and quality of forests?
   A. Carrying capacity
   B. Forestry
   C. Silviculture
   D. Manipulative management

4. _____ What is the term used for a practice that prevents wildlife species from declining by minimizing external factors such as human interaction or urban development?
   A. Carrying capacity
   B. Silviculture
   C. Manipulative management
   D. Custodial management

5. _____ What is term used to describe the ability of the habitat to support wildlife?
   A. Carrying capacity
   B. Silviculture
   C. Manipulative management
   D. Custodial management

6. What does BTU stand for? What is a BTU? _____
   _____
   _____

7. _____ Which type of management would the following scenario be classified as: A national park discourages visitors from feeding wild animals.
   A. Manipulative management
   B. Custodial management

8. _____ Which type of management would the following scenario be classified as: Approving a certain number of hunting licenses each year for a predatory species.
   A. Manipulative management
   B. Custodial management

Copyright Goodheart-Willcox Co., Inc.
May not be reproduced or posted to a publicly accessible website.

9. How many nuclear power plants are there in the United States, and how much electrical energy do they produce each year? _____
_____
_____

10. Describe the difference between renewable and nonrenewable energy resources. Provide an example of each from the text. _____
_____
_____

11. Why is wildlife management necessary? _____
_____
_____

Name _____ Date _____ Class _____

 **Activity 10.5A**

Forest Products in Our Lives: A Scavenger Hunt

After doing some initial research and review of Lesson 10.5 to brush up on your knowledge of forest products, get out a sheet of paper and, if available, a camera. On your teacher's mark, race around your classroom (or hallway, school, and/or outdoor areas surrounding your building, if allowed) to collect images and/or record as many forest products in your surroundings as possible! Once the agreed-upon time allotment is up, come back together with your classmates to compare images/lists of products you found. Whoever has the most wins. Happy hunting!

Name _____ Date _____ Class _____

LESSON 10.6 — United States Forest Regions

Lesson Review

Carefully study the lesson and then answer the following questions.

1. _____ are found in Alaska and across Canada, where temperatures remain very low throughout the year and about half of the precipitation is snow.
   _____

2. In _____, the winter months are very mild, the days are usually 12 hours or more, and there is an average of more than 78 inches of rainfall annually.
   _____

3. A(n) _____ is any forest that receives more than 100 inches of rainfall each year.
   _____

4. Boreal forests are also called _____.
   _____

5. _____ are located primarily in the Eastern United States, experience four distinct seasons each year, and exist in moderate climates.
   _____

6. Name four tree species found in boreal forests:
   A. _____   C. _____
   B. _____   D. _____

7. Name four species found in temperate forests:
   A. _____   C. _____
   B. _____   D. _____

8. Describe the characteristics of the primary forest regions of the US.
   A. _____
   B. _____
   C. _____

9. Identify the three major forest regions in the United States.
   A. _____
   B. _____
   C. _____

10. Explain why different species thrive in different regions. _____
    _____
    _____

Name _____ Date _____ Class _____

 **Activity 10.6A**

American Tree Profile

Now that we've learned about the major forest regions found within the United States, let's see if we can apply that to our surroundings! Take a walk with your class outside of your school building, or even at home, and collect four different leaves from tree species in your area. Do some research to identify the tree from which each leaf came from and include the following information with each leaf after attaching it to a sheet of paper:

- Standard name and scientific name of the tree
- Forest region(s) found in
- Average age
- Average size
- Type of soil preferred
- Preferred climate
- Identifiable traits (leaf shape, coloring, bark, etc.)

Name _____ Date _____ Class _____

Forests and Forest Products

Lesson Review

Carefully study the lesson and then answer the following questions.

1. _____ is an oil distilled from pine trees' sap and is used in paints and medications.
   _____

2. A(n) _____ is an area where hazardous chemicals pose immediate danger to humans and the environment.
   _____

3. _____ is a thick sticky substance derived from tree sap and used to waterproof rope, the sails on sailing ships, and wooden-hulled ships.
   _____

4. Trees that are too small to make into lumber are used instead to make _____.
   _____

5. Scientists have created a special species of _____ tree that can break down common environmental pollutants.
   _____

6. _____ is a versatile construction item used as the foundation for flooring, roofing, and even as a structural load-bearing unit in building homes.
   _____

7. A thicker version of tar is called _____ and was used as a waterproofing agent to caulk the seams in wooden ships.
   _____

8. _____ is wood used for lumber.
   _____

9. What four pollutants are commonly found at superfund sites?

   A. _____

   B. _____

   C. _____

   D. _____

10. _____ is the process that uses carbon, oxygen, and hydrogen to create plant sugars and starches essential for plant growth.
    _____

11. Explain why forests and trees are important to people and the planet. _____
    _____

Name _____ Date _____ Class _____

 Activity 10.7A

Tasty Tree Treats

Not only are trees the "lungs of the Earth," playing incredibly vital roles in our world, but they also provide some of the tastiest and most nutritious food products in our diets! From maple syrup to citrus, nuts, and olives, trees provide many of our favorite things to eat. Find a food product that comes from trees to present to your classmates. What does that tree look like? What are the stages the tree goes through to produce your chosen product, and what does it look like at each of those stages? How long does it take to produce that product? Does the product need to go through any sort of processing before arriving to a grocery store? Where are those trees grown and why?

To make this even more fun, turn each presentation into a snackable moment by providing samples of each item presented! (Be aware that some students in your class may have allergies, particularly to tree nuts. Please get a teacher's approval before bringing a treat in to share!)

Name _____ Date _____ Class _____

LESSON 10.8
Forest Management Practices

Lesson Review
Carefully study the lesson and then answer the following questions.

1. Trees with shallow or damaged roots may experience ____, which occurs when the trees fall over in high winds.
_____

2. _____ What is the term for the parts of the tree left in the woods after a tree is harvested, like the leftover branches, twigs, leaves, bark, and other bits?
   A. Inventory
   B. Windthrow
   C. Slash
   D. Cuttings

3. _____ Before cutting timber, a landowner should conduct a forest ____, noting the tree species, soil type(s), and wildlife in a forest.
   A. Inventory
   B. Sustainable forest management
   C. Stream buffer zone
   D. Appraisal

4. _____ cuts down all the trees in a given area, resulting in ample open space that is ready for reseeding or transplanting young trees.
_____

5. _____ What is the term for techniques and practices that protect soil and water resources in forests?
   A. Inventory
   B. Sustainable forest management
   C. Stream buffer zone
   D. Appraisal

6. Loggers mark ____, or areas next to streams where logging operations are prohibited.
_____

7. _____ What is the term for an estimate of the worth of something?
   A. Inventory
   B. Log Deck
   C. Selection Cutting
   D. Appraisal

8. Explain the difference between *selection cutting* and *shelterwood cutting*. _____
_____
_____

9. The ____ is a central location in a timber cutting operation where log skidders bring cut timber for trimming and stacking.
_____

10. _____ define property boundaries to ensure that logging operations do not infringe on others' property.
_____

Name _____

11. What is the purpose of forest management? _____
_____
_____

12. What three items are necessary for a forest management plan? _____
_____
_____

13. Explain how prescribed fires and wildfires differ, and how both are a part of forest management. _____
_____
_____

Name _____ Date _____ Class _____

Activity 10.8A

Forest Management Plan in Action

Using the information provided in Lesson 10.8, go to a local forested area to survey the state and health of that forest. Create a forest management plan based on what you see! Take photos to support your findings and present them to your classmates.

Name _____ Date _____ Class _____

LESSON 10.9 Anatomy of a Tree

Lesson Review

Carefully study the lesson and then use the word bank to fill in the blanks in each question.

1. _____ The living part of the stem is called the _____.
2. _____ A secondary stem that branches from the main stem and contains specialized growing cells on the tip of the branch is called a _____.
3. _____ The edge of a leaf is called a(n) _____.
4. _____ _____ have many seeds in the center instead of just one seed.
5. _____ The tree's _____ protects its main stem and is made of old phloem tissue that has died and been replaced with new phloem tissue.
6. _____ A _____ has many seeds that are grouped in the core of the fruit.
7. _____ Xylem and phloem from previous seasons die and become _____, which is solid and resistant to decay and insect damage.
8. _____ A _____ is a fruit with an edible seed enclosed in a hard shell coating.
9. _____ A _____ is a fleshy fruit with one seed in the center.
10. _____ At the top of each growing stem is a specialized area of fast-growing cells called the _____.

A. heartwood
B. sapwood
C. drupe
D. nut
E. bark
F. berries
G. margin
H. apical meristem
I. pome
J. lateral meristem

Carefully study the lesson and then answer the following questions.

11. What purpose do stems serve for trees? _____
_____
_____

12. Why are roots important to trees? _____
_____
_____

13. How are tree flowers pollinated? _____
_____
_____

Copyright Goodheart-Willcox Co., Inc.
May not be reproduced or posted to a publicly accessible website.

Name _____ Date _____ Class _____

 **Activity 10.9A**

Pollination Process

Using the information in Lesson 10.9, go to a local forested area in the vicinity of your school or home and identify the reproductive organ of three tree species you find. Take a photo of each.

You can also look around the produce in your home or at a local supermarket and see if you can identify a drupe, a berry, a pome, and a nut, taking a photo of each.

Compare your findings with your classmates!

Name _____ Date _____ Class _____

LESSON 10.10: Commercially Important American Trees

Lesson Review

Carefully study the lesson and then answer each question.

1. Tree species such as oak, apple, and walnut trees are also known as ____.
   _____

2. ____ is a measure used extensively in the lumber industry.
   _____

3. ____ are trees with seeds that do not have a protective coating, such as pines and spruces.
   _____

4. ____ are trees that have seeds with a covering, like oak, apple, and walnut species.
   _____

5. Tree species such as pines, spruce, and other softwoods are also known as ____.
   _____

6. Explain how trees are classified as hardwood or softwood. _____
   _____
   _____

7. _____ Which tree species is native to the Southeastern United States and is the principal lumber tree for that region?
   A. Fraser fir
   B. Loblolly pine
   C. Coast redwood
   D. Maple

8. _____ Which tree species is resistant to decay, insects, and disease but needs fog for healthy growth?
   A. Fraser fir
   B. Ash
   C. Coast redwood
   D. Yellow poplar

9. _____ Which tree species is also called a "tulip tree," is fast-growing, and native to the Eastern United States?
   A. Walnut
   B. Loblolly pine
   C. Coast redwood
   D. Yellow poplar

10. _____ Which tree species has wood that is shock-resistant, making it an excellent choice for items like baseball bats and boat paddles?
    A. Fraser fir
    B. Ash
    C. Aspen
    D. Walnut

Copyright Goodheart-Willcox Co., Inc.
May not be reproduced or posted to a publicly accessible website.

11. _____ Which tree species is native to the Appalachian Mountains and is grown primarily for Christmas trees?
    A. Fraser fir
    B. Loblolly pine
    C. California black oak
    D. Yellow poplar

12. Identify 10 species of commercially important hardwood trees in the United States.

    A. _____
    B. _____
    C. _____
    D. _____
    E. _____
    F. _____
    G. _____
    H. _____
    I. _____
    J. _____

13. Identify 8 species of commercially important softwood trees in the United States.

    A. _____
    B. _____
    C. _____
    D. _____
    E. _____
    F. _____
    G. _____
    H. _____

Name _____ Date _____ Class _____

 **Activity 10.10A**

Commercial Tree Memory Game

Choose a partner and get out some index cards. (You can thank a tree for those!) Write the name of an important commercial tree species we learned about in Lesson 10.10 on each card. On the opposite side of the card, include the following information for each species:

- Hardwood or softwood
- Location grown
- Commercial uses
- Visibly identifiable characteristics
- Interesting fact (like nicknames, specific growing conditions needed, etc.)

Once your card deck is complete, take turns going through the cards with your partner. One partner will read the description of the tree species while the other one uses that information to guess the correct name. Continue by switching back and forth until time is up, and use these to help you study later!

Name _____ Date _____ Class _____

LESSON 10.11 / Natural Resources Management Practices in Agriculture

Lesson Review

Carefully study the lesson and then answer each question.

1. _____ is a method by which a seedbed is tilled without disturbing the soil on either side.
   _____

2. _____ is the practice of planting crops on an alternating basis.
   _____

3. A(n) _____ is similar to a cover crop but uses a mixture of native plants and wildflowers.
   _____

4. Farmers and ranchers use _____ between crop harvests to prevent soil erosion and maintain soil moisture.
   _____

5. _____ leaves the soil in place from year to year, which allows a layer of organic matter to build up.
   _____

6. A(n) _____ is a narrow swath of plants that put space between the crop and other areas or streams and ponds.
   _____

7. The three positives of no-till are:
   A. _____
   B. _____
   C. _____

8. _____ occurs when farmers and ranchers move livestock from one pasture to another to prevent overgrazing.
   _____

9. _____ are crop varieties developed by scientists that can withstand dry weather and need less irrigation.
   _____

10. Explain why it is important to conserve water and soil resources. _____
    _____
    _____

11. List at least four common soil and water conservation practices in agriculture. _____
    _____
    _____

Name _____ Date _____ Class _____

Activity 10.11A

Erosion with Our Own Eyes

In Lesson 10.11, we focused on the importance of natural resource conservation and the ways agriculturalists implement conservation techniques to protect water and soil resources. Let's see if we can expand our understanding of erosion with our own eyes through this fun activity!

To begin, gather the following materials:
- Various ground substances like clay, wood chips, soil, sand, and gravel
- Water
- Plastic bottles
- Plastic cups

Once you have your materials, cut your plastic bottles in half and fill with a variety of soils, like soil and rocks, soil and wood chips, or soil with sand. Set each mixture to be on a slant, so that there is more of the fill mixture on one side than the other.

Next, pour water on each soil mixture at the highest point. As the water trickles down, be sure to collect any run-off using the plastic cups.

Examine the water in each cup and discuss which mixture has the most soil in the water, or most erosion. Which mixture kept the most soil? Why?

Name _____ Date _____ Class _____

Wildlife and Game Management

Lesson Review

Carefully study the lesson and then answer each question.

1. _____ are those species of animals that may be legally harvested through hunting, trapping, or fishing.
   _____

2. _____ removes dirt and helps to apply a waxy substance secreted from a gland near the bird's tail to the feathers, which protects them and makes them waterproof.
   _____

3. *Ornithology* is the study of _____.
   _____

4. Amphibians go through a life cycle where the young grow from eggs into adults, with significant body changes. This process is called _____.
   _____

5. Most birds are _____, meaning they can perch on surfaces such as limbs and twigs.
   _____

6. _____ are mammals that can use arms and legs, or front and rear legs, for moving from place to place.
   _____

7. Describe *radio telemetry*. _____
   _____
   _____

8. Describe the methods that wildlife biologists use to study wildlife. _____
   _____
   _____
   _____
   _____

Name _____

Carefully study the lesson and then use the word bank to match the term to the correct definition.

9. _____ This animal is native to the Eastern United States, is omnivorous, and is classified as a near threatened species due to habitat loss.

10. _____ This animal came to the United States from Asia and Africa, can reproduce quickly, and poses a serious problem in feral populations.

11. _____ The most widely distributed species of its kind, this animal can adapt to a wide range of habitats and is an herbivorous ruminant.

12. _____ This animal is native to North America and prefers to live in mixed hardwood-conifer forests, although it will forage for food in open pastures and fields.

13. _____ This animal is native to North America, is the most widely distributed species of its kind, typically inhabit forests, and primarily eats vegetation.

14. _____ This animal inhabits farmland, woodland, and wetlands, and was introduced to the United States in 1773 from Asia and Europe. Hunting this animal is a popular sport: In the last 30 years the US population has decreased.

A. Wild turkey
B. Common pheasant
C. White-tailed deer
D. Northern bobwhite
E. Wild boar
F. American black bear

Name _____ Date _____ Class _____

Activity 10.12A

The Cornell Lab of Ornithology: Live Cams

Visit The Cornell Lab of Ornithology's website to view their Live Bird Cams for a virtual window into the lives of birds! This site can be found through a quick internet search.

Take a moment to explore the live cams and pick your favorite one. What do you see? What kinds of birds do you see? How do they behave? What sounds do they make? Spend some time observing and taking notes to share with the class.

Name _____ Date _____ Class _____

Safe Agricultural Work Practices

Lesson Review

Carefully study the lesson and then answer the following questions.

1. _____ The primary goal in all agricultural workplaces and laboratories should be to maintain a(n) _____.
    A. equipment safety manual
    B. safe work environment
    C. material safety data sheet
    D. clean work surface

2. What is critical to ensuring safe operation of equipment and completing work correctly? _____
   _____

3. _____ *True or False?* Over time, hearing loss can be gained back.

4. What are three reasons that an indoor workspace might require ventilation? _____
   _____
   _____
   _____

5. How should chemicals be properly stored? _____
   _____

Match the description with the correct injury type.

6. _____ Jumping from a piece of scaffolding onto the ground and breaking a bone   A. Contact injury
7. _____ Pulling a back muscle from moving boxes from one conveyor belt to another   B. Impact injury
8. _____ Getting your foot ran over by a forklift   C. Nonimpact injury
9. _____ Cutting your hand on a piece of sheet metal
10. _____ Running into a trailer hitch while loading lumber

11. Identify five areas that may be protected by personal protective equipment (PPE). _____
    _____
    _____
    _____
    _____
    _____

12. _____ A(n) _____ is necessary to have when you're planning events to take place in any facility.
    A. evacuation map
    B. first-aid kit
    C. chemical storage cabinet
    D. safety plan

Copyright Goodheart-Willcox Co., Inc.
May not be reproduced or posted to a publicly accessible website.

Name _____ Date _____ Class _____

Activity 11.1A

Safety in Practice

Imagine that you are working in a laboratory facility in your school. This may be a greenhouse, a livestock barn, an ag mechanics shop, etc. Use complete sentences to answer the following questions.

Chosen Facility: _____

1. Describe at least five potential hazards or accidents that are present in this environment when a class full of students are present.

   _____
   _____
   _____
   _____
   _____
   _____
   _____
   _____

2. Identify at least five examples of PPE that could be necessary in this facility and identify a type of injury that the PPE could offer protection against.

   _____
   _____
   _____
   _____
   _____
   _____
   _____
   _____

3. Create a safety poster that could be posted in your chosen facility that displays the potential hazards and proper PPE that should be used in that facility.

   _____
   _____
   _____
   _____
   _____
   _____
   _____
   _____

Name _____ Date _____ Class _____

Simple Machines and Common Tools

Lesson Review

Carefully study the lesson and then answer the following questions.

1. _____ Which of the following is *not* an example of a cutting tool?
   A. Tin snips
   B. Crosscut handsaw
   C. Utility knife
   D. Wire strippers

2. Which category of hand tools is frequently used to hold materials so that they can be modified? _____
   _____
   _____

3. _____ *True or False?* Stationary power tools are commonly used for mass producing materials.

4. Explain the difference between an electric drill and a pneumatic drill. _____
   _____
   _____

Match each example with the simple machine that it represents.

5. _____ Elevator
6. _____ Prybar
7. _____ Scissors
8. _____ Door knob
9. _____ Cement mixer
10. _____ Runaway truck ramp
11. _____ Hand truck
12. _____ Window blinds
13. _____ Post hole borer
14. _____ Pizza cutter

A. Inclined plane
B. Lever
C. Pulley
D. Screw
E. Wedge
F. Wheel and axle

Copyright Goodheart-Willcox Co., Inc.
May not be reproduced or posted to a publicly accessible website.

Name _____ Date _____ Class _____

 **Activity 11.2A**

Simple Machines in Agriculture

Choose five agricultural areas from the list below. Identify five tools that are used in each of the selected areas. Write down the name of the tool and the simple machine(s) it is comprised of. Explain how each tool is used in its respective agricultural area.

Agricultural Areas

Animal Production	Food Processing	Turfgrass/Landscaping
Aquaculture	Forestry	Veterinary Science
Crop Production	Horticulture	Wildlife Management
Equipment Maintenance and Repair	Metal Fabrication	Woodworking

Agricultural Area #1: _____

Tools

1. _____

    A. Simple Machine(s): _____

    B. How it's used: _____

    _____

    _____

2. _____

    A. Simple Machine(s): _____

    B. How it's used: _____

    _____

    _____

3. _____

    A. Simple Machine(s): _____

    B. How it's used: _____

    _____

    _____

4. _____

    A. Simple Machine(s): _____

Name _____

   B. How it's used: _____
_____
_____

5. _____

   A. Simple Machine(s): _____
_____

   B. How it's used: _____
_____
_____

Agricultural Area #2: _____

Tools

1. _____

   A. Simple Machine(s): _____
_____

   B. How it's used: _____
_____
_____

2. _____

   A. Simple Machine(s): _____
_____

   B. How it's used: _____
_____
_____

3. _____

   A. Simple Machine(s): _____
_____

   B. How it's used: _____
_____
_____

4. _____

   A. Simple Machine(s): _____
_____

   B. How it's used: _____
_____
_____

5. _____

   A. Simple Machine(s): _____

   B. How it's used: _____

Agricultural Area #3: _____

Tools

1. _____

   A. Simple Machine(s): _____

   B. How it's used: _____

2. _____

   A. Simple Machine(s): _____

   B. How it's used: _____

3. _____

   A. Simple Machine(s): _____

   B. How it's used: _____

4. _____

   A. Simple Machine(s): _____

   B. How it's used: _____

5. _____

   A. Simple Machine(s): _____

Name _____

B. How it's used: _____
_____
_____

Agricultural Area #4: _____

Tools

1. _____

   A. Simple Machine(s): _____
   _____

   B. How it's used: _____
   _____
   _____

2. _____

   A. Simple Machine(s): _____
   _____

   B. How it's used: _____
   _____
   _____

3. _____

   A. Simple Machine(s): _____
   _____

   B. How it's used: _____
   _____
   _____

4. _____

   A. Simple Machine(s): _____
   _____

   B. How it's used: _____
   _____
   _____

5. _____

   A. Simple Machine(s): _____
   _____

   B. How it's used: _____
   _____
   _____

Agricultural Area #5: _____

Tools

1. _____

   A. Simple Machine(s): _____
   _____

   B. How it's used: _____
   _____
   _____

2. _____

   A. Simple Machine(s): _____
   _____

   B. How it's used: _____
   _____
   _____

3. _____

   A. Simple Machine(s): _____
   _____

   B. How it's used: _____
   _____
   _____

4. _____

   A. Simple Machine(s): _____
   _____

   B. How it's used: _____
   _____
   _____

5. _____

   A. Simple Machine(s): _____
   _____

   B. How it's used: _____
   _____
   _____

LESSON 11.3 Measurement and Layout

Lesson Review

Carefully study the lesson and then answer the following questions.

1. _____ The stability and precision of a structure is checked using _____ measurements.
   A. general
   B. angular
   C. reflective
   D. pressure

2. _____ Which of the following is *not* a basic unit of measurement in the metric system?
   A. Liter
   B. Gram
   C. Acre
   D. Meter

3. _____ _____ calculations are beneficial in figuring out the amount of material that is necessary to cover a structure.
   A. Precision
   B. Surface area
   C. Mass
   D. Layout

4. _____ *True or False?* The level of accuracy required is dependent upon the type of work that is being finished.

5. _____ _____ is the measurable period during which a process occurs.
   A. Time
   B. Speed
   C. Velocity
   D. Impact

6. _____ Which of the following is *not* measured by volume?
   A. Concrete
   B. Fill dirt
   C. Sod
   D. Mulch

7. _____ The boiling point of water is _____ degrees Fahrenheit and _____ degrees Celsius.
   A. 100; 212
   B. 0; 100
   C. 212; 100
   D. 212; 37

8. Why were there great discrepancies in measuring using the original form of measurement in ancient China?
   _____
   _____

9. How many countries have *not* adopted the metric system of measurement? _____

10. Explain why one bushel of brown-top millet is the same size as one bushel of soybeans, but weighs a different amount.
    _____
    _____

Name _____ Date _____ Class _____

 Activity 11.3A

Metric System Scavenger Hunt

Look around your classroom, in your bookbag, and any location that your teacher gives you permission to search to find objects that are a specific mass, length, or volume. Try to find objects that are as close as possible to the provided mass, length, or volume. Once you've identified those items, calculate the difference between the desired measurement and the actual measurement.

Materials
- Calculator
- Electronic Scale or Triple Beam Balance
- Meter Stick

Mass Measurements

Desired Mass	Object	Actual Measurement (g)	Difference (g)
10.0 g			
343.2 g			
794.6 g			
1276.0 g			
2021.5 g			

Total Difference (g): _____

Length Measurements

Desired Length	Object	Actual Measurement (cm)	Difference (cm)
1.4 cm			
21.5 cm			
30.4 cm			
50.8 cm			
91.5 cm			

Total Difference (cm): _____

182 Exploring Agriculture Food and Natural Resources Lab Workbook

Name _____

Volume Measurements

Desired Volume	Object	Actual Measurement (mL)	Difference (mL)
119.2 mL			
236.6 mL			
354.9 mL			
499.8 mL			
709.0 mL			

Total Difference (mL): _____

1. What is the metric unit of mass? _____

2. What is the metric unit of length? _____

3. What is the metric unit of volume? _____

4. What are the metric prefixes in order from smallest to largest? _____
_____

Name _____ Date _____ Class _____

LESSON 11.4
Common Fasteners and Materials

Lesson Review

Carefully study the lesson and then answer the following questions.

1. _____ Which of the following is not a common grouping of fasteners?
   A. Bolts
   B. Nails
   C. Nuts
   D. Screws

2. The three parts of a nail are:

   A. _____   C. _____

   B. _____

3. _____ *True or False?* A 16d nail is shorter in diameter and length than a 10d nail.

4. _____ The three basic head types of a screw are _____.
   A. flat, star, and square
   B. flat, oval, and round
   C. oval, round, and star
   D. oval, round, and square

5. What is the difference between a screw and a bolt? _____
   _____

6. _____ *True or False?* Most lumber is produced from hardwoods because it is easier to work with.

7. What is the difference between plywood and oriented strand board (OSB)? _____
   _____

Match the description with the correct metal type.

8. _____ able to be drawn into wire   A. Ductile

9. _____ can be shaped or formed   B. Fusible

10. _____ easily melted or fused   C. Malleable

11. What alloy is produced by combining copper with zinc? _____
    _____

12. What alloy is produced by combining copper with tin? _____
    _____

13. Which category of plastics is often combined with fiberglass to increase the strength? _____
    _____

14. What are the four components of concrete? _____
    _____

Name _____ Date _____ Class _____

 **Activity 11.4A**

Design-A-Fastener

Create your own fastener using the guidelines provided below. You'll design a nail, a screw, and a bolt. Once you've chosen what your fasteners will look like, you need to create a model of each design. The models can be made from poster board, modeling clay, or any other type of material that you choose. Models should be made to scale.

Fastener #1: Nail
Design your nail to have a nail head, shank, and tip.

    Material that the nail is made from: _____

    Application (interior/exterior): _____

    Shank type (ring, spiral, barbed, or smooth): _____

    Driving method (hand tool/power tool): _____

    End use (roofing, masonry, concrete, etc.): _____

    Diameter: _____

    Shank length: _____

Fastener #2: Screw
Design your screw to have a screw head, shank, and tip.

    Thread type (fine/coarse): _____

    Shank thread (fully/partial): _____

    Head type (flat/oval/round): _____

    Driving slot (flat/Phillips/square/star): _____

    Driving method (hand tool/power tool): _____

    Shank length: _____

Fastener #3: Bolt
Design your bolt to have a head and shank. Also design a nut that fits over your bolt.

    Shank thread (fully/partial): _____

    Requires washers (yes/no): _____

    Head shape: _____

    Bolt grade: _____

    Shank length: _____

Name _____ Date _____ Class _____

LESSON 11.5
Irrigation and Plumbing

Lesson Review

Carefully study the lesson and then answer the following questions.

1. _____ The process of applying controlled amounts of water to land to aid in producing crops is _____.
   A. fertilization
   B. irrigation
   C. aquification
   D. preservation

2. What type of water is found in lakes, rivers, and reservoirs? _____

3. _____ Approximately _____ percent of the US population rely on well water.
   A. 2
   B. 7
   C. 14
   D. 22

Match the surface irrigation type to the correct definition.

4. _____ Water is distributed across the soil surface through sprinkler systems

5. _____ Water is distributed across the soil surface by the flow of gravity

6. _____ Water is applied immediately above the soil surface in small drops or streams

   A. Drip
   B. Flooding
   C. Sprinkler

7. What is the difference between surface irrigation and subsurface irrigation? _____
   _____

8. _____ The primary line in a water supply system is a(n) _____.
   A. aquifer
   B. reservoir
   C. septic tank
   D. water main

9. _____ Water used more than once before it moves back into the natural water cycle is called _____.
   A. reclaimed water
   B. recycled water
   C. repurposed water
   D. reused water

Match the parts of a sanitary drainage system with their definition.

10. _____ Device that prevents the passage of sewer gas into the building or structure

11. _____ Carries liquid waste from a sink, bathtub, or shower to a drain

12. _____ A fitting with a removable plug to allow clear access for removing obstructions

13. _____ Carries water and fecal matter from a toilet to a drain or sewer

14. _____ Allows air into the drainage system

15. _____ A vertical line of water, soil, or vent pipes that extends through one or multiple stories in a structure

   A. Cleanout
   B. Fixture drainage trap
   C. Soil pipe
   D. Stack
   E. Vent pipe
   F. Waste pipe

Name _____ Date _____ Class _____

 **Activity 11.5A**

Create Your Own Irrigation System

In this activity, you'll build a simple irrigation system with plastic cups and straws. The system will distribute water by using only a gravitational pull.

Materials
- Clear plastic cups
- Drinking straws
- Scissors
- Modeling clay
- Water

Instructions

1. Take one cup. Using the scissors, cut x-shaped slits in the top half of the cup, on opposite sides of the cup.

2. Take two more cups. Cut a single x-shaped slit in the bottom half of each cup. Make sure the slits on the two cups are lower than the slit on the original cup.

3. Connect the system by placing the ends of the drinking straws into the slits. Seal the slits off by using modeling clay to keep the water from leaking.

4. Slowly pour water into your main cup (this will act as your water source).

   A. What happens when the water in the main cup reaches "straw level"?

   _____
   _____

5. Add one more cup off of the main cup. Try to make the x-shaped slit at a different height than the other two.

   A. Are all 3 cups able to receive the same amount of water when you pour water into the main cup? Why or why not?

   _____
   _____

6. Add one cup off of a secondary cup. Make sure that the x-shaped slit is below the slit that you created to receive water from the main cup.

   A. Is it possible to get the same amount of water in each cup? Why or why not?

   _____
   _____

7. Attach two straws together by using modeling clay to seal the area where they attach. Add a new cup to a secondary cup.

   A. Does the water move into the new cup? What can you learn about real-world irrigation systems from this?

   _____
   _____

Name _____ Date _____ Class _____

LESSON 11.6 Fundamentals of Electricity

Lesson Review

Carefully study the lesson and then answer the following questions.

1. _____ _____ current can be transmitted at high voltages over great distances.
   A. Alternating
   B. Direct
   C. Parallel
   D. Series

2. Lead-acid and lithium-ion batteries are examples of _____ batteries.
   _____

3. What is electromagnetic induction? _____
   _____
   _____
   _____

4. _____ The frequency of change in a cycle is called a(n) _____.
   A. amp
   B. hertz
   C. ohm
   D. volt

Match the electrical unit of measurement with the correct definition.

5. _____ The pressure that moves electrons through a conductor
6. _____ The rate of flow of the electrical current
7. _____ The opposition of the electrical flow
8. _____ The rate at which the work of an electrical circuit is completed

A. Amperage
B. Resistance
C. Voltage
D. Wattage

9. Identify four examples of natural resources utilized by power plants to power generators that send electricity to supply lines. _____
   _____
   _____
   _____

10. What is the difference between a series circuit and a parallel circuit? _____
    _____
    _____
    _____

Name _____ Date _____ Class _____

Activity 11.6A

Resistance in Electricity

The resistance of a material is dependent on four factors: the makeup of the material, the size of the material, the length of the material, and the temperature of the material. In this activity, you will measure the resistance of copper wire in varying conditions.

Materials

- 3 pieces of 10-gauge copper wire cut to 24″ (61 cm)
- 3 pieces of 12-gauge copper wire cut to 24″ (61 cm)
- 3 pieces of 14-gauge copper wire cut to 24″ (61 cm)
- Ohmmeter

Measure and record the resistance of each scenario listed below by using an ohmmeter. Attach the red lead to one end of the wire and the black lead to the other end. Make sure that your ohmmeter is set to the resistance (ohm) setting.

1. 14-gauge wire at room temperature: _____ ohms
2. 14-gauge wire in direct sunlight for 1 hour: _____ ohms
3. 14-gauge wire in a freezer for 1 hour: _____ ohms
4. 12-gauge wire at room temperature: _____ ohms
5. 12-gauge wire in direct sunlight for 1 hour: _____ ohms
6. 12-gauge wire in a freezer for 1 hour: _____ ohms
7. 10-gauge wire at room temperature: _____ ohms
8. 10-gauge wire in direct sunlight for 1 hour: _____ ohms
9. 10-gauge wire in a freezer for 1 hour: _____ ohms

Name _____ Date _____ Class _____

Small Engine Basics and Maintenance

LESSON 11.7

Lesson Review

Carefully study the lesson and then answer the following questions.

1. Identify three pieces of equipment that utilize small engines. _____
   _____
   _____

Match the engine part to the correct definition.

2. _____ Is made up of the piston, piston rings, piston pin, and connecting rod
3. _____ Produces the proper mixture of fuel and air to ignite and operate an engine
4. _____ Converts the up and down motion of pistons into a circular motion
5. _____ Comprised of the intake and exhaust valves and the camshaft
6. _____ A single piece of machined cast iron or aluminum
7. _____ A heavy, revolving wheel that increases the engine's momentum

   A. Carburetor
   B. Crankshaft
   C. Engine block
   D. Flywheel
   E. Piston assembly
   F. Valve Train

8. _____ Which of the following is *not* a part of a small engine fuel system?
   A. Air filter
   B. Fuel lines
   C. Spark plug
   D. Supply tank

9. What are the three parts of a small engine's cooling system? _____
   _____
   _____

10. What tool is used to check the gap on a spark plug? _____
    _____

Name _____ Date _____ Class _____

 **Activity 11.7A**

Small Engine Measurements

The questions are designed to acquaint you with the operational features of a small OHV gasoline engine. Fill in the blanks below carefully as you disassemble your engine.

Materials
- Small OHV gasoline engine (Briggs & Stratton OHV 130000 preferred, but this activity can be modified to fit any small engine)

1. Engine Manufacturer: _____

   Model #: _____

   Type #: _____

   Code #: _____

   Date of engine manufacturing? _____

2. What is the oil capacity of the engine? _____ ounces

3. What are the minimum API and SAE Standards for this engine?

   API Classification: _____

   SAE Viscosity Rating: _____

4. What are the torque specifications for the following components? Be sure to list the unit of measure.

   Cylinder Head: _____

   Crankcase Cover: _____

   Flywheel Nut: _____

   Connecting Rod: _____

   Rocker Ball Lock Nut: _____

5. What is the compression ratio of this engine? _____

6. Does your engine pass a spark test? _____

7. Using a feeler gauge, what is the armature air-gap?

   Found armature air-gap: _____

   Recommended armature air-gap range: _____

8. Check the spark plug gap with a wire feeler gauge:

   Found spark plug gap: _____

   Recommended spark plug gap: _____

9. Measure the valve clearance for each valve.

   Intake Valve

   Found Clearance: _____

   Recommended Clearance Range: _____

   Exhaust Valve

   Found Clearance: _____

   Recommended Clearance Range: _____

10. Measure the crankshaft journal. Record both found and reject sizes.

    Magneto Journal

    Found size: _____

    Reject size: _____

    PTO Journal

    Found size: _____

    Reject size: _____

    Crankpin Journal

    Found size: _____

    Reject size: _____

11. Measure the piston ring end gap. Record both found and reject sizes.

    Top Compression Ring

    Found size: _____

    Reject size: _____

    Middle Oil Wiper Ring

    Found size: _____

    Reject size: _____

    Oil Control Ring

    Found size: _____

    Reject size: _____

12. Measure and record the following information. Record both found and reject sizes.

    Intake Valve Stem Diameter

    Found size: _____

    Reject size: _____

    Exhaust Valve Stem Diameter

    Found size: _____

    Reject size: _____

Name _____

Camshaft Bearing Bore Diameter

Found size: _____

Reject size: _____

13. What is the crankshaft end play for this engine?

    Crankshaft end play: _____

14. Measure and record both the found and reject size for the camshaft, cam gear, and cam lobes.

    Camshaft Mag Journal

    Standard size: _____

    Reject size: _____

    Camshaft PTO Journal

    Standard size: _____

    Reject size: _____

15. What is the cylinder bore diameter? _____

16. What is the stroke of this engine? _____

17. How many cubic inches of displacement does this engine have? _____

18. Find the standard and reject size for the parts listed below.

    Cylinder Bore Diameter

    Standard size: _____

    Reject size: _____

    Cylinder Bore Out of Round

    Standard size: _____

    Reject size: _____

    Piston Pin Diameter

    Standard size: _____

    Reject size: _____

    Piston Pin Bore Diameter

    Standard size: _____

    Reject size: _____

Name _____ Date _____ Class _____

LESSON 11.8
Project Planning, Design, and Construction

Lesson Review

Carefully study the lesson and then answer the following questions.

1. _____ A project budget makes sure you consider the _____ of material during the planning stage.
   A. availability
   B. cost
   C. longevity
   D. size

2. Identify four potential costs that need to be factored into a project budget. _____
   _____
   _____

3. _____ *True or False?* Computer-assisted design (CAD) programs are used to complete three-dimensional and four-dimensional drawings.

Match the construction process to the correct definition.

4. _____ Assembly
5. _____ Finish
6. _____ Material selection
7. _____ Preparation

A. Based on location, use, availability, quality, and cost
B. Welding, fastening, and/or constructing the materials
C. Protects the project from corrosion and provides an attractive appearance
D. Organizing, marking, and/or cutting to prepare for assembly

8. Explain the difference between a rip cut and a crosscut. _____
   _____
   _____

9. Explain the difference in nominal and actual lumber dimensions. _____
   _____
   _____
   _____

10. _____ Which grade of softwood is best used for shelving?
    A. Construction
    B. No. 2 common
    C. Select
    D. Utility

Name _____ Date _____ Class _____

Activity 11.8A

Bill of Materials

1. Find the unit price.

   A. $120 wages paid for 8 hours of work: _____

   B. $176 wages paid for 8 hours of work: _____

   C. $100 per box of 400 sticks: _____

   D. $110 per box of 400 sticks: _____

   E. $25.50 per box of 850 nails: _____

   F. $22.50 per box of 850 nails: _____

2. A box of 750 nails costs $37.50.

   A. What is the cost of one nail? _____

   B. Joe needs 75 nails and wants 10 extra in case he bends some. What will his total cost be, if he buys the nails individually for the same unit price as the box price? _____

3. A box of 750 nails costs $35.50.

   A. What is the cost of one nail? _____

   B. If the box of nails weighs 20 pounds, what is the weight of one nail? _____

   C. Jenny needs 75 nails and wants 10 extra in case she bends some. What will her total cost be, if she buys the nails individually for the same unit price as the box price? _____

   D. What is the total weight of 85 nails? _____

4. Jose is making a jewelry box that needs two hinges. A package of six hinges costs $7.98.

   A. What is the cost of one hinge? _____

   B. What is the cost of two hinges? _____

5. Julie is making a jewelry box that needs two hinges. A package of six hinges costs $8.29.

   A. What is the cost of one hinge? _____

   B. What is the cost of two hinges? _____

6. Annette took her trailer to a machinist to get a new axle machined. Her total bill was $335.00 for materials and labor. She was billed $35.00 for materials. If it took the machinist 5 hours to complete the repair and make the axle, what did they charge Annette per hour for labor? _____

7. Matt took his trailer to a machinist to get a new axle machined. His total bill was $379.00 for materials and labor. He was billed $41.25 for materials. If it took the machinist 5 hours to complete the repair and make the axle, what did they charge Matt per hour for labor? _____

8. Sheet metal costs $96.00 per sheet and measures 4 by 8 feet. There are 32 square feet in one sheet. What is the cost per square foot? _____

9. Sheet metal costs $99.00 per sheet and measures 4 by 8 feet. There are 32 square feet in one sheet.

   A. What is the cost per square foot? _____

   B. There are 144 square inches in one square foot. What is the cost per square inch? _____

10. Sue found sales at two different hardware stores. She could buy 5 buckets of paint for $24.96 or 7 buckets of paint for $35.19. What is the difference in the price per bucket of paint from each sale? _____
_____

11. Stan found sales at two different hardware stores. He could buy 7 boxes of nails for $32.96 or 9 boxes of nails for $39.19. What is the difference in the price per box of nails from each sale? _____
_____

Name _____ Date _____ Class _____

LESSON 12.1 — What Is Marketing?

Lesson Review

Carefully study the lesson and then answer the following questions.

1. What is the purpose of marketing? _____
   _____
   _____

2. Identify four steps of marketing a product. _____
   _____
   _____
   _____

3. _____ *True or False?* Marketing nonfood products includes more steps than products that are grown, harvested, or graded.

Match the type of agricultural market with the correct description.

4. _____ Places more emphasis on providing entertainment and seasonal attractions
5. _____ Creates personal connections and mutual benefits between farmers, shoppers, and the local community
6. _____ Provides opportunities for farmers to spend less time on marketing and more time on developing a high-quality product

A. Community Supported Agriculture
B. Farmers markets
C. Recreational marketing

7. Integrators can be directly involved with various aspects of producing a product. Viewpoints vary about whether integrators are positive or negative influences on the market. Which characterization do you think is more accurate? Use the information you learned from the chapter to justify your answer. _____
   _____
   _____

8. Why do farmers earn more for their produce when selling to a retailer, rather than a wholesaler? _____
   _____
   _____

9. How are wholesalers able to provide large quantities to consumers? _____
   _____
   _____

10. _____ All processing facilities set _____ for the crops that they receive.
    A. base prices
    B. profit margins
    C. quantities
    D. standards

Copyright Goodheart-Willcox Co., Inc.
May not be reproduced or posted to a publicly accessible website.

11. _____ *True or False?* Recreational marketing approaches make up a large percentage of the money spent on agricultural products in the United States.

12. _____ Businesses are able to build and maintain consumer relationships if they recognize the importance of _____.
    A. friendship
    B. marketing
    C. packaging
    D. selling

Agricultural Markets

Create a presentation showcasing agricultural businesses in your community and/or state. Identify at least 5 businesses that produce an agricultural product.

1. Business: _____

   Product: _____

2. Business: _____

   Product: _____

3. Business: _____

   Product: _____

4. Business: _____

   Product: _____

5. Business: _____

   Product: _____

Develop a map of the area that your businesses represent (community, county, state, etc.) and create a presentation (poster, brochure, digital, or another appropriate choice) that displays each of the businesses, what they offer, and where they're located.

Name _____ Date _____ Class _____

LESSON 12.2 USDA Standards and Grades

Lesson Review

Carefully study the lesson and then answer the following questions.

1. _____ *True or False?* Food label regulations and monitoring are the responsibility of state-level agencies.

2. Identify seven commodity categories that have been identified by the USDA. _____
   _____

3. _____ The USDA inspects products for freshness, cleanliness, and _____.
   A. grade
   B. marketability
   C. purity
   D. shelf life

4. _____ *True or False?* The standards outlined by the USDA refer to the product, as well as the facilities where the product is handled.

5. Why do large-volume buyers utilize the USDA grading system for their business transactions? _____
   _____

6. Identify four aspects or characteristics that commodities may be graded on. _____
   _____

7. Identify three commodities that USDA quality grade marks are found on. _____
   _____

8. _____ Which is the correct order of recognized beef grades from lowest quality to highest quality.
   A. Choice, Prime, Select, Standard
   B. Prime, Select, Choice, Standard
   C. Standard, Select, Choice, Prime
   D. Select, Standard, Choice, Prime

9. _____ *True or False?* Growing organic products using fewer chemicals will always produce higher yields as a result.

10. _____ Which of the following is *not* an approved organic label?
    A. 100% Non-GMO
    B. Made with Organic
    C. Organic
    D. 100% Organic

Name _____ Date _____ Class _____

 **Activity 12.2A**

Egg Grading

Grade eggs based on their exterior quality by using the guidelines outlined by the USDA.

Materials

Dozen eggs (freshly laid if possible; store-bought eggs purchased at different times to give varying ages if not)

Exterior Grading Procedure

1. Observe the exterior of the egg closely and carefully. DO NOT TOUCH THE EGGS!
2. While observing the egg's exterior, look closely at the following factors:
   A. Shell stains or discolorations
   B. Overall shape of the shell
   C. Overall texture of the shell
   D. Thin spots
   E. Cleanliness of the shell

The table below gives you more information what to look at for each of these characteristics.

3. Select a USDA grade (AA, A, B, or REJECT) for each egg.
   A. AA is the best quality; B is the lowest quality.
   B. Eggs that are not suitable for human use in any form are classified as REJECT.

FACTOR	GRADE AA OR A	GRADE B	GRADE REJECT
Stains or Discolorations	Clean, may have small specks, stains, or cage marks that do not affect the overall appearance of the egg; may show traces of processing oil	Slight or moderate stains on less than 1/32nd of the shell in one area **OR** Slight or moderate scatter stains on less than 1/16th of the shell	Prominent stains **OR** Slight or moderate stains that cover more than 1/32nd if in one area or 1/16th of the shell if scattered
Overall Shape	Appears to have usual egg shape	Unusually or decidedly misshapen	
Overall Texture	May have rough areas, small calcium deposits, but do not materially affect shape or strength of shell	Extremely rough areas that may be faulty in soundness or strength; large calcium deposits.	
Shell Thickness	Free of thin spots	May have pronounced thin spots	
Cleanliness	None	None	Adhering dirt or other materials (1.0mm in area or greater)

Exterior Grades

Egg #1: _____

Egg #2: _____

Egg #3: _____

Egg #4: _____

Egg #5: _____

Egg #6: _____

Egg #7: _____

Egg #8: _____

Egg #9: _____

Egg #10: _____

Egg #11: _____

Egg #12: _____

Name _____ Date _____ Class _____

LESSON 12.3
Determining Price Points

Lesson Review

Carefully study the lesson and then answer the following questions.

1. Describe the relationship between supply and demand. _____
   _____
   _____

2. Identify four factors that can affect supply. _____
   _____
   _____

3. _____ Demand can be affected by interest rates, consumer opinion, and _____.
   A. crop yields
   B. product costs
   C. surplus
   D. wage rates

4. Why would a restaurant print "market price" in their menu instead of a precise dollar amount? _____
   _____
   _____

5. What are the two main drivers that cause market prices to fluctuate? _____
   _____
   _____

Match the factor to the related description.

6. _____ Maintain a higher market price

7. _____ Limit supply for manufacturers, slowing down production

8. _____ May affect how careful consumers are with how they spend their money

A. Employment and wages
B. Natural disasters; world events
C. Luxury items

9. _____ The retail price that keeps a product in high demand is a(n) _____.
   A. demand price
   B. market price
   C. price point
   D. wholesale price

10. Identify three factors to consider when selecting the best price point. _____
    _____
    _____

Copyright Goodheart-Willcox Co., Inc.
May not be reproduced or posted to a publicly accessible website.

Name _____ Date _____ Class _____

Activity 12.3A

Determining and Comparing Price Points

Determine the price point needed to meet the projected profit.

Cost	Selling Price	Profit
$135		10%
$405		15%
$639		25%
$268		32%
$57		40%

Compare prices from various retailers.

Item:	Retailer #1: Price:	Retailer #2: Price:	Retailer #3: Price:
12-pack of Sharpie Markers			
Miracle-Gro 6 qt. Indoor Potting Mix			
32-pack AA Energizer Batteries			
UNO Card Game			
Grip-Rite 16d 3-1/2" Common Nails – 1 lb. Box			

Did these prices vary more or less than you thought they might? Why do you think this is the case?

_____

_____

_____

204 Exploring Agriculture Food and Natural Resources Lab Workbook

Copyright Goodheart-Willcox Co., Inc.
May not be reproduced or posted to a publicly accessible website.

Name _____ Date _____ Class _____

LESSON 12.4 Marketing Strategies and Plans

Lesson Review
Carefully study the lesson and then answer the following questions.

1. What is the main difference between marketing and selling? _____
_____
_____

2. _____ The overall plan that a business develops to reach consumers and make an effort to turn them into customers is a _____.
   A. market objective
   B. marketing plan
   C. market research
   D. marketing strategy

3. What are the four Ps of marketing? _____
_____
_____

4. _____ A marketing plan helps a business focus their efforts on _____.
   A. maximizing customer base
   B. maximizing profits
   C. minimizing customer base
   D. minimizing profits

5. _____ Which of the following is *not* a purpose of a marketing plan?
   A. Obtain target audience preferences
   B. Grow customer base
   C. Increase the amount of sales
   D. Set clear objectives

6. Identify two methods that can be used to conduct market research. _____
_____
_____

Copyright Goodheart-Willcox Co., Inc.
May not be reproduced or posted to a publicly accessible website.

 **Activity 12.4A**

Guess the Audience

It is important to understand the audience that your business is trying to reach. By tailoring your marketing strategies and plans to fit your audience, you can ensure that the message will be better received.

One student will work to guess the audience that the remainder of their classmates are portraying. The designated student will leave the classroom while the remaining students decide who they want to be as an audience (i.e.: grandparents, kindergarten students, sports fans, etc.). Once the student returns, they will ask probing questions (i.e.: What is your favorite food? What is your favorite thing to do in your free time?) A member of the audience will respond with an answer that is tailored towards your chosen audience. Once the student can correctly guess who their target audience is, other students can volunteer to guess a new audience.

Name _____ Date _____ Class _____

Promoting and Advertising Agricultural Products

Lesson Review

Carefully study the lesson and then answer the following questions.

1. Identify two examples of a promotion effort. _____
   _____
   _____

2. Identify two examples of a promotion concept. _____
   _____
   _____

3. Identify two examples of a promotion item. _____
   _____
   _____

4. _____ Handing out free samples to customers is a good way to increase _____.
   A. brand awareness
   B. customer traffic
   C. informed consumers
   D. marketing strategies

5. _____ Which is the most effective free way to promote a business?
   A. Business cards and branded promotional materials
   B. Twitter campaign
   C. Website
   D. Word of mouth

6. Why would a business participate in a charity sponsorship? _____
   _____
   _____

7. _____ *True or False?* Word of mouth advertising is the fastest way to keep customers up to date on product availability and current business sales.

8. _____ The main goal of advertising is to entice customers to _____.
   A. buy a product
   B. follow the business on all social media platforms
   C. tell their friends about the business
   D. visit the store

9. _____ *True or False?* Having a solid plan helps to ensure that businesses are spending their money on effective advertisements that reach their intended audience.

10. Identify the 5 components of an advertising plan. _____
    _____
    _____

Name _____ Date _____ Class _____

Activity 12.5A

Promoting and Advertising Agricultural Products

Choose an agricultural product that you want to advertise. The product can be as large as a combine or as small as a bottle of motor oil. Complete the questions below and design an advertisement for your product.

1. What is your product? _____
   _____

2. Where can it be purchased? _____
   _____
   _____

3. Who is your target audience? (Be specific!) _____
   _____
   _____

4. What makes this product better than the competition? _____
   _____
   _____

5. What message are you trying to portray? _____
   _____
   _____

6. How will you advertise your product? Which advertising platform will you use to reach your target audience?
   _____
   _____

Depending on your chosen advertising platform, create an advertisement for your product. The advertisement should include a photo of your product, brand name, and additional information about your product (i.e.: price, description, specials, varieties, etc.). Imagine you have a budget of $750 to advertise your product. How will you spend it?

   In the space below, explain the overall cost of your advertisement and provide as much detail as possible.

_____
_____
_____
_____
_____
_____
_____
_____

Name _____ Date _____ Class _____

LESSON 12.6 Technology in Marketing and Agribusiness

Lesson Review

Carefully study the lesson and then answer the following questions.

1. _____ The US Bureau of Economic Analysis states that online sales and services make up _____ of all US services and goods sold.
   A. 2.1%
   B. 9.6%
   C. 12.6%
   D. 19.1%

2. _____ A _____ is the path raw materials follow as they move from one company to another.
   A. flow chart
   B. processing diagram
   C. market analysis
   D. supply chain

Match the term to the correct description.

3. _____ Method that creates correspondence between customers and agribusinesses without the need for human input

4. _____ Computer software programs that track websites and specific customer-viewed products

5. _____ Strategy used to show advertisements from webpages visited previously

6. _____ Method used to send newsletters, coupons, and product and service information

7. _____ Strategy used to get customers to stay longer on a website

8. _____ Computer software program that is used to find information on the internet

A. Automation
B. Analytics
C. Conversion optimization
D. Email
E. Remarketing
F. Search engine

9. Identify two ways that a company might utilize technology to maintain their records. _____
_____
_____

10. Viewpoints vary about whether new technology has had a positive or negative influence on agriculture and agribusiness. Do you think the overall effect has been more positive or negative? Use the information you learned from the chapter to justify your answer. _____
_____
_____
_____
_____

Copyright Goodheart-Willcox Co., Inc.
May not be reproduced or posted to a publicly accessible website.

Name _____ Date _____ Class _____

Activity 12.6A

Tracking Inventory

You are employed by Dill's Ace Hardware Stores, a local business that provides agricultural goods within your community. Using the information provided, complete the inventory sheets and questions.

1. In the table below, find the total amount of each product that was sold for the week.

Dill's Ace Hardware Store Sales						
Item Name	Mon. Sales	Tue. Sales	Wed. Sales	Thurs. Sales	Fri. Sales	Total Sales
RV & Marine Antifreeze	1	4	6	5	4	
Milwaukee Brushless Hammer Drill Kit	2	3	0	5	3	
1" Filtrete 1000 Air Filter	5	3	2	1	3	
First Alert Smoke Alarm	5	10	4	8	6	
Ace 50 ft. All-Season Garden Hose	10	5	6	4	5	
Miracle-Gro Moisture Control Potting Mix 1 cu. ft.	4	3	4	3	0	
Milwaukee 25' Compact Wide Blade Tape Measure	5	4	3	7	10	
Fiet Electric LED Smart Bulb 3-pack	2	5	1	0	3	
Duracell AA Batteries – 8-pack	25	10	5	7	16	
Rubbermaid Brute 32-gallon Trash Can	10	11	10	8	5	
Ace 10W-30 Motor Oil	10	15	25	5	18	
Ortho Home Defense Insecticide	15	20	15	15	15	
Milwaukee Torque Lock Locking Pliers	12	4	0	10	3	
Kaytee Birders' Blend Wild Bird Food – 8 lb.	5	5	5	5	5	
Ace Winterizer Weed and Feed	4	10	0	0	0	
Igloo IMX 70 Quart Cooler	1	0	4	8	3	
Craftsman 35 pc. Right Angle Rachet Set	4	6	7	10	3	
10' × 10' Pop-Up Canopy	4	6	7	0	12	
DeWalt 10" Circular Saw Blade	6	15	4	0	0	
Little Giant Velocity 22' Aluminum Ladder	0	6	0	3	0	

Name _____

2. Using the Beginning Inventory table below, fill in Column A on the Dill's Ace Hardware Store Ending Inventory at the end of this exercise.

Dill's Ace Hardware Store Beginning Inventory (Monday Morning)	
Item Name	Beginning Inventory
RV & Marine Antifreeze	20
Milwaukee Brushless Hammer Drill Kit	20
1" Filtrete 1000 Air Filter	35
First Alert Smoke Alarm	50
Ace 50 ft. All-Season Garden Hose	45
Miracle-Gro Moisture Control Potting Mix 1 cu. ft.	28
Milwaukee 25' Compact Wide Blade Tape Measure	55
Feit Electric LED Smart Bulb 3-pack	15
Duracell AA Batteries – 8-pack	75
Rubbermaid Brute 32-gallon Trash Can	80
Ace 10W-30 Motor Oil	100
Ortho Home Defense Insecticide	110
Milwaukee Torque Lock Locking Pliers	46
Kaytee Birders' Blend Wild Bird Food – 8 lb.	25
Ace Winterizer Weed and Feed	25
Igloo IMX 70 Quart Cooler	30
Craftsman 35 pc. Right Angle Rachet Set	37
10' × 10' Pop-Up Canopy	38
DeWalt 10" Circular Saw Blade	25
Little Giant Velocity 22 Aluminum Ladder	19

3. Next, complete Column B on the Dill's Ace Hardware Store Ending Inventory sheet at the end of this exercise by recording the numbers you totaled in Question #1.
4. Complete Column C on the Dill's Ace Hardware Store Ending Inventory sheet at the end of this exercise by calculating the total items received (below).

Dill's Ace Hardware Store Order Sheet

Item Name	Per Package	Quantity Ordered	Total Items Received
RV & Marine Antifreeze	72	1	
Milwaukee Brushless Hammer Drill Kit	72	1	
1" Filtrete 1000 Air Filter	72	1	
First Alert Smoke Alarm	24	2	
Ace 50 ft. All-Season Garden Hose	24	4	
Miracle-Gro Moisture Control Potting Mix 1 cu. ft.	24	2	
Milwaukee 25' Compact Wide Blade Tape Measure	36	1	
Feit Electric LED Smart Bulb 3-pack	0	0	
Duracell AA Batteries – 8-pack	100	1	
Rubbermaid Brute 32-gallon Trash Can	0	0	
Ace 10W-30 Motor Oil	12	3	
Ortho Home Defense Insecticide	12	4	
Milwaukee Torque Lock Locking Pliers	12	1	
Kaytee Birders' Blend Wild Bird Food – 8 lb.	12	1	
Ace Winterizer Weed and Feed	24	1	
Igloo IMX 70 quart Cooler	0	0	
Craftsman 35 pc. Right Angle Rachet Set	0	0	
10' × 10' Pop-Up Canopy	12	1	
DeWalt 10" Circular Saw Blade	24	2	
Little Giant Velocity 22' Aluminum Ladder	0	0	

Name _____

5. Calculate the ending inventory totals for each item listed on the Dill's Ace Hardware Store Ending Inventory. Subtract the total sales from your beginning inventory and add any orders that arrived during the week.

Dill's Ace Hardware Store Ending Inventory (Friday Afternoon)

Item Name	Beginning Inventory (Mon. Morning)	Total Sales (-)	Store Orders (+)	Ending Inventory
RV & Marine Antifreeze				
Milwaukee Brushless Hammer Drill Kit				
1" Filtrete 1000 Air Filter				
First Alert Smoke Alarm				
Ace 50 ft. All-Season Garden Hose				
Miracle-Gro Moisture Control Potting Mix 1 cu. ft.				
Milwaukee 25' Compact Wide Blade Tape Measure				
Feit Electric LED Smart Bulb 3-pack				
Duracell AA Batteries – 8-pack				
Rubbermaid Brute 32-gallon Trash Can				
Ace 10W-30 Motor Oil				
Ortho Home Defense Insecticide				
Milwaukee Torque Lock Locking Pliers				
Kaytee Birders' Blend Wild Bird Food – 8 lb.				
Ace Winterizer Weed and Feed				
Igloo IMX 70 quart Cooler				
Craftsman 35 pc. Right Angle Rachet Set				
10' × 10' Pop-Up Canopy				
DeWalt 10" Circular Saw Blade				
Little Giant Velocity 22 Aluminum Ladder				

Questions

1. Which items sold the most during the week? _____
   _____

2. Which items sold the least during the week? _____
   _____

3. The store has set an inventory policy to maintain a quantity of at least 25 of each item in stock. Using the ending inventory amounts, which items will need to be re-ordered for next week? _____
   _____
   _____